Challenges of Modernization and Governance in South Korea

Jae-Jung Suh · Mikyoung Kim
Editors

Challenges of Modernization and Governance in South Korea

The Sinking of the Sewol and Its Causes

palgrave
macmillan

Editors
Jae-Jung Suh
Department of Politics and International Studies
International Christian University
Tokyo, Japan

Mikyoung Kim
Chair, Human Rights Research Committee
International Political Science Association (IPSA)
Quebec, Canada

ISBN 978-981-10-4022-1 ISBN 978-981-10-4023-8 (eBook)
DOI 10.1007/978-981-10-4023-8

Library of Congress Control Number: 2017936034

Cover image: clipartdotcom

Printed on acid-free paper

This Palgrave Macmillan imprint is published by Springer Nature
The registered company is Springer Nature Singapore Pte Ltd.
The registered company address is: 152 Beach Road, #21-01/04 Gateway East, Singapore 189721, Singapore

We dedicate this book to the memory of the Sewol victims and the resilience of their families

Acknowledgements

I was in Korea for sabbatical leave when the Sewol sank. It was a surreal experience. I was one of the millions of people who were fixated on the TV screen watching the ferry sinking. The media was live broadcasting the multiple deaths and we were the helpless collective witnessing the live burial. The high-school students who were supposed to be enjoying their school fieldtrip to Jeju Island were instead drowning under the water right before our eyes. The whole thing about it was absolutely, totally, and unequivocally wrong.

I kept on running into the same sight of the people fixated on the TV screen in the neighborhood restaurants, coffee shops, buses, and trains. I could not escape from their blank eyes and pale faces no matter where I went. And I stopped going out. My mind refused to accept the senseless deaths of hundreds of innocent people, and my body resisted being forced to witness their live burials. No matter how radically the world changed in this age of instant information-sharing, certain aspects of our lives seemed to deserve respect and sanctification. I grounded myself in a small, dark room that I was renting in the metropolitan Seoul area, and did not come out for many days. I had to shut myself out of Korean society for self-preservation. I was afraid of going mad. And yet I probably did.

I could no longer tell what was real and what was not, what to trust and what not to. I had missed my homeland and wanted to return to the place of nostalgic, warm memories. I had wanted to get reconnected to the society which had grown and developed while I had been gone for 25 years. I had been proud of Korea. Then the Sewol tragedy, all of a sudden, confronted me with an unexpected question. What had happened to it?

I wanted to find an answer to what went wrong, and why and how it happened. And I talked to Jae-Jung about compiling a volume on the sinking of the Sewol and South Korea. That would be our way of making sense out of the absurd and giving meaning to the sacrifice of the victims and their grieving families. Little did I or he know that the conversation would be the beginning of a long journey that led to many unexpected encounters. On the journey, we kept running into many colleagues who were besieged with, and troubled by, the same question, whether they were based in Korea, the USA, or Australia. Some of them had already written about the Sewol; others had done research on issues that would help us make sense of the nonsensical; some had been working with the victims' families in the city of Ansan; and many others had been doing what they as academics could to cope with the tragedy. When we proposed collaborating on a volume that would trace the causes of the Sewol tragedy to deeper socio-political ills of Korean society, most of them readily accepted. Along the journey that took much longer than any of us had anticipated, not only did they bear with us without complaints but also thanked us, repeatedly, for keeping the project up. It is now my turn to thank them. This volume would not have been possible without their contributions, nor could I have come this far without their patience.

One of the first things we did to critically examine the multiple causes of the Sewol tragedy was to organize a roundtable at the International Studies Association held in New Orleans in February 2015. Taehyun Nam deserves special recognition for organizing it under the auspices of the Association of Korean Political Studies. I would also like to thank Yoonkyung Lee and Jong-sung You for their participation in the panel and the volume.

I do not know how to thank another group of people, some of whom I encountered along this journey: the families of the Sewol victims. Here let me share a story of a colleague. The colleague developed indigestion and signs of an ulcer after helplessly watching the live broadcast of hundreds of the innocent being drowned. It was only when the families began to demand to know the true cause of the death of their loved ones that she realized she was not that helpless after all. She could do something as mundane as signing a petition in support of the families and the thought gave her a sense of relief as well as a glimmer of hope. She said, "I was able to get out of the hollow feeling of powerlessness thanks to the courage and perseverance of the families." The families who were hit the hardest by the tragedy stood up and waged a campaign to make Korea safer lest others should suffer the same loss. I was able to get out of the dark room where I was hiding because I found strength in them. I could start and complete the volume because of their inspiration.

Also, I would like to acknowledge, with a tremendous sense of gratitude, the assistance provided by Human Rights Foundation Saram. Its staff helped some of the contributors meet with Sewol families and secure copyright clearance. On behalf of all the authors, I thank them for their selfless assistance, without which the volume would be much less than it is now.

I must extend my sincere apology to the contributors. I could not keep my promise of writing a chapter on the South Korean government's defamation lawsuit against the *Sankei Shimbun*. I wanted to compare and analyze the post-article and post-lawsuit reactions in the Korean and Japanese news media. Upon my return to Japan from Korea in February 2015, I had to face many professional challenges at my workplace. It involved my new boss, a.k.a. Mr. Rightwing, his allies at the university and my fights to protect junior female colleagues. This situation did not allow me to engage in productive research activities for the past 1,126 days. My chapter could not be delivered. I am sorry.

Despite the heartless and mindless fights that were going on for more than three years, the warm support from my co-editor, Jae-Jung, and other devoted chapter contributors kept me above the water. I am most grateful for JJ's selfless work during my sudden disappearnace due to house raid, arrest and 12-day detention at the criminal charge filed by my former employer in Japan, Hiroshima City University. The university fired me, a tenured faculty, on the day of my release at the Prosecution's dropping of the charge. While I was going through a living hell, JJ remained calm, kind and very supportive. He is the best colleague one can have. The Palgrave editors, Sara Crowley-Vigneau, Nina Li, and Connie Li, were more than generous with their time and efforts to make this volume strong.

The volume is dedicated to the memory of the Sewol victims. Lest their deaths should be forgotten, the names of the victims are imprinted on the covers. Should we fail to heed their warning, South Korea too may sink someday. As we hope Koreans learn from the ferry's sinking, so we are mindful that the sinking of the Sewol may portend the sinking of South Korea.

It is dedicated also to the love and resilience that their families have shown in keeping their memory alive and making Korean society safer. May the deceased rest in peace. May their families find peace.

Mikyoung Kim

Contents

Editors And Contributors

About the editors

Jae-Jung Suh is Professor at International Christian University in Tokyo, Japan. He has served as Associate Professor and Director of Korea Studies at SAIS, Johns Hopkins University and Assistant Professor in the Department of Government at Cornell University, as well as on the Presidential Commission on Policy Planning (Republic of Korea). An expert on US–Korea relations, US policy toward Asia, international relations of East Asia, international security, and IR theory, he has authored and edited numerous journal articles and books, including *Power, Interest and Identity in Military Alliances* (2007); *Rethinking Security in East Asia: Identity, Power and Efficiency* (2004); *Truth and Reconciliation in the Republic of Korea: Between the Present and Future of the Korean Wars* (2012); and *Origins of North Korea's Juche: Colonialism, War, and Development* (2012), among others.

Mikyoung Kim is IPSA Human Rights Research Committee Chair (http://rc26.ipsa.org) and President of Association of Korean Political Studies (www.akps.org). She is the author of many refereed articles and books on memory, reconciliation and human rights in East Asia. She also has contributed many newspaper oponion pieces in English, Korean and Japanese. She was a two-term member of the ROK National Unification Council (2011–14). Her book, Routledge Handbook of Memory and

Reconciliation in East Asia, won the best book award by the ROK Ministry of Education in 2016. She is currently preparing a lawsuit to recover her tenured faculty position at Hiroshima City University in Japan. The university filed a criminal charge against her on March 6, 2016, leading to her detention for 12 days. Even though the charge was dropped by the Prosecution Office, the university fired her on the day of the release.

Contributors

Su-Dol Kang Korea University, Sejong, South Korea

Hyeon Jung Lee Seoul National University, Seoul, South Korea

Hyunok Lee Yonsei University, Wonjoo, South Korea

Yoonkyung Lee University of Toronto, Toronto, Canada

John Lie University of California, Berkeley, USA

Seungsook Moon Department of Sociology, Vassar College, Poughkeepsie, USA

Taehyun Nam Salisbury University, Salisbury, USA

K. S. Park Korea University, Seoul, South Korea

Youn Min Park Australian National University, Canberra, Australia

Jae-Jung Suh International Christian University, Mitaka, Japan

Jong-sung You Australian National University, Canberra, Australia

List of Figures

List of Tables

CHAPTER 1

Introduction: Sinking the Sewol, Drowning Korea? Compressed Modernization and Compounded Risks

Jae-Jung Suh

The Sewol ferry carrying 476 passengers including a group of high school students on a field trip to Jeju Island capsized on April 16, 2014, and sank to the bottom of the sea off Korea's southern coast. Most of the crew, including the captain, were rescued by the Korean Coast Guard. Some of the passengers, who happened to be on the deck or escaped soon after the capsizing, were saved by fishing boats and commercial vessels that came before the Republic of Korea (ROK) Coast Guard or Navy. However, 304 passengers were trapped inside and drowned.

The ship's sinking may seem an unfortunate accident, the operation to save the passengers a heroic drama enacted in the seas, and the passengers' death its tragic ending. The whole sequence that unfolded before the nation's eyes makes nothing short of a tragedy. Once the surface is scratched, however, a more complicated picture emerges. The Sewol sank under the weight of the state that was entangled with the maritime

J.-J. Suh (✉)
International Christian University, Mitaka, Japan

J.-J. Suh and M. Kim (eds.), *Challenges of Modernization and Governance in South Korea*, DOI 10.1007/978-981-10-4023-8_1

industry's self-serving interests. The rescue operation was weighed down by an absentee state that delegated its responsibility to a private salvage firm. And yet when questions arose about the state's responsibility, the state was not shy about mobilizing its resources to evade and deny responsibility. The whole incident served as a prism through which the power of Korea's strong state was refracted differently depending on issue areas. The state was small and neoliberal in some areas but large and strong in others.

A careful analysis of the picture, furthermore, reveals more troubling, and perhaps more deeply rooted, ills that inflict Korea. The entanglement between the state and the maritime industry is part of a larger totalitarian pattern the roots of which can be traced back at least to the 1960s. The education that produced the students who listened to the crew's instruction only to their demise remains a legacy of an authoritarian discipline imposed by the authoritarian state. Recent "democratic" reforms have produced the ironical effect of eviscerating students even more. Added to this is the state's suppression of the media and freedom of expression, generating pressure on people to remain silent and docile to power. The pressure comes with an addiction: decades of single-minded pursuit of development have addicted all organizations, from the state to businesses, to a profit-driven worldview that willingly sacrifices everything else. The addiction has been interdigitated with Korea's modernization in a compressed time frame. The nation, so preoccupied with rushed development, has grafted one sociopolitical hazard on top of another without ever pausing to reflect critically upon them, much less address them. Korea's compressed modernization has instead kept these hazards under pressure. And the Sewol disaster is a symptom and a warning.

The first part of this introductory chapter provides an overview of the Sewol disaster in terms of the three stages of its unfolding: the Sewol's sinking, rescue failures, and the aftermath. It analyzes different roles played by the state in the Sewol incident with a view toward demonstrating that the power of the state in Korea is exercised unevenly in the three stages. The second part then takes the analysis deeper to suggest that the Sewol disaster is a symptom of more serious maladies with which Korea is afflicted. The country's modernization has generated a risk society that such scholars as Beck and Giddens note is an inalienable part of the modernization process.[1] Because it has compressed its modernization, it has grafted hazards onto others without pausing to reflect seriously on, much less resolve, any. Korea in the twenty-first century is a risk society *par excellence* that keeps these compounded hazards under pressure. Some of the hazards momentarily revealed their destructive potential in the Sewol disaster—perhaps as a warning that should the nation go on further without exercising self-reflexivity, it may invite upon itself more disastrous consequences.

I. Sinking of the Sewol, Drowning of the Passengers[2]

The whole tragedy can be divided into three periods: before, during, and after the sinking of the Sewol. The Korean state played a central role in each, sowing the seed of a tragedy, growing it to its full potential, and making the outcome more tragic with denials. Before the Sewol's final voyage, it had colluded with the shipping industry in a symbiotic relationship that benefited officials and businessmen at the cost of citizens' safety. During the ferry's sinking, the state disappeared, shirking its responsibility to save the passengers from the very danger it had helped create. After the sinking, it reared its almighty power to hide its wrongdoings and silence the victims' families and critics. This section chronicles its actions in the three periods, and the following one places them in the historical context of Korea's compressed development that has compounded risks in the society.

The Sewol Sinks Under the Weight of the Diminishing State

The ROK Coast Guard concluded on April 17, 2014 that an "unreasonably sudden turn" to starboard, made between 8:48am and 8:49am, was the cause of the capsizing. The Automatic Identification System (AIS)

---

[1] For the concept of risk society, see Beck, Ulrich. *Risk Society: Towards a New Modernity.* London; Newbury Park, Calif.: Sage Publications, 1992; and Giddens, Anthony. "Risk and Responsibility." *The Modern Law Review* 62, no. 1 (1999): 1–10 (Ulrich 1992; Giddens Anthony 1999).

[2] This section is a revised and updated version of Jae-Jung Suh, "The Failure of the South Korean National Security State: The Sewol Tragedy in the Age of Neoliberalism," *The Asia-Pacific Journal*, Vol. 12, Issue 40, No. 1, October 6, (2014) (Suh 2014). http://apjjf.org/2014/12/40/Jae-Jung-Suh/4195.html.

data that kept the ship's trajectory until its sinking seemed to confirm the sudden turn. But the Sewol's trajectory still remains unconfirmed, and the exact cause of its sinking unknown.[3] The Coast Guard's conclusion itself leaves many critical questions unanswered.

First, why did the third mate at the helm have to make a turn so sudden and steep as to capsize the ship? At the trial on June 10, the third mate testified that she instructed the quartermaster to turn the ship by five degrees in order to avoid colliding with "a ship" that was approaching. She testified in court that a ship "emerged" from the opposite side and "she was watching the radar and the front while listening to the radio in order to avoid a collision." Her testimony is corroborated by a video recording by a commercial ship that was passing by the Sewol at the time of the accident: it shows an object moving towards the Sewol. Also, AIS data restored from the Mokpo AIS Station—the Sewol's own AIS was turned off at 8:48 am for an unknown reason—shows a trail left by an independent moving object.

Quartermaster Cho stated in a TV interview on April 19 that he turned the helm as ordered by the third mate, but when he did, the ship turned more than usual. When the surviving passengers told reporters that they felt a collision shock in the front side of the ship, their remark initially prompted speculation that the ship might have hit a reef or rock under water. But when the complete trajectory of the Sewol was released after a long delay, it suggested a different story. It showed that not only did the ferry turn by the five degrees as ordered by the third mate, but also gradually changed course almost 180° as if it had been pushed by an object traveling with great momentum in the opposite direction.

What caused the Sewol crew to make a sudden turn remains unexplained. What forced the ship to change its course and move in the opposite direction from the previous course likewise remains a mystery.

Second, why did the Sewol capsize when it changed course? Investigations revealed that the ship had been modified to accommodate

---

[3] Kim Ji-Young, a documentary filmmaker who has followed the incident to produce a documentary, has convincingly pointed out many problems in the official AIS data submitted by the Ministry of Oceans and Fisheries. http://www.hanitv.com/index.php?category=52596&document_srl=199687&page=2 and http://www.hanitv.com/index.php?category=52596&document_srl=206077&page=2 His suspicion was that the officially released data had been adjusted to move the ferry's trajectory approximately 200 m north of its true voyage. 문형구, "청해진해운이 국정원에 보낸 의문의 AIS좌표," 미디어오늘, 2016년5월18일. http://www.mediatoday.co.kr/?mod=news&act=articleView&idxno=130032.

more passengers than would be safe. Added to the overcrowding was cargo overloading. The ship's operators loaded twice as much as regulations would allow, and apparently did not secure the cargo as per safety guidelines. To accommodate the overloading, the crew removed water from the ballast, creating perfect conditions for capsizing. The ferry had excessively heavy cargoes that moved around and too many passengers who were told to stay put while too little water remained in the ballast to stabilize the ship.

This raises a host of questions. Why was the Chonghaejin Marine Co., Ltd., the Sewol's owner, allowed to add more floors than the safety law allowed? How could the operators overload the ship without being caught? How could they remove ballast water to the ship's peril? Why were none of these violations caught or stopped before the Sewol embarked on its ill-fated journey? These questions lead one to the shadowy relationship between shippers, the shipping industry organization and the regulators—*haefia*, a Korean syllogism that concatenates *hae* meaning sea with *fia* from Mafia: "sea Mafia." It is an iron triangle of the sea that thrives in the age of neoliberalism.

The Sewol had been built and operated in Japan for almost 18 years without any accidents until 2012 when it was retired and sold to Chonghaejin Marine Co., Ltd., which had a monopoly on the lucrative Inchon-Jeju line. The Korean owner bought the ferry after the Lee Myung-Bak administration extended passenger ships' life spans from 25 to 30 years by changing the relevant law's implementation rules, thus allowing the Sewol another 5 years of life in Korea. The neoliberal administration that vigorously pursued deregulation justified the extension on the grounds that it would help the Korean shipping industry save $20 million per year in operating costs and become profitable. It thus unambiguously placed industry profit before passengers' safety.

If the Ministry of Land and Sea Management took active steps to deregulate, another wing of the government provided Chonghaejin with the cash needed for it to take advantage of the deregulation. Chonghaejin itself had been established by absorbing the Semo Marine Transportation after 200 billion Won (roughly $200 million) of its debt was forgiven. The Korea Development Bank, a wholly state-owned bank that finances major industrial projects, then loaned $10 million, an amount that almost matched the $12 million that Chonghaejin paid for the Sewol. Taking advantage of deregulation and the government's generosity, Chonghaejin added two floors in order to maximize the number of passengers it could accommodate. It also expanded the Sewol's cargo space.

There was still one more obstacle to overcome before Chonghaejin could turn deregulation and the policy loan into a real profit. It had to pass the safety inspection before it could launch the Sewol. It passed this without difficulty. While the additions of the floors undermined the ship's stability by adding more weight at the top and thus endangered passengers' safety, the modification was inspected and approved by inspectors from the Korean Register of Shipping (KRS), a private entity that is responsible for the inspection and registration of ships. Also when it inspected over 200 safety features of the Sewol in February 2014, it approved all with a "satisfactory" rating. Prosecutors investigating the process of inspection and approval have discovered that government oversight of the KRS, which performs the safety inspection on its behalf, has been lax. A cause might lie in the fact that government regulators frequently find employment at the KRS after retirement.

Chonghaejin's greed did not stop there. It took advantage of a loophole in the government's safety regulations to routinely overload the Sewol. In Korea's coastal shipping industry, shippers' safety practices are monitored and inspected by the Korea Shipping Association (KSA), an industry organization that represents the interests of about 2000 members engaged in coastal shipping. In a flagrant case of self-regulation, its headquarters is responsible for "safety guidance" and "implementation of safety measures" while its branch offices are tasked with offering "guidance for passenger ferry's safe operation" and inspecting the number of passengers and the amount of cargo aboard a ship. The Marine Transportation Law creates the position of vessel safety operators to guide and oversee the shipping businesses' safety practices, but the safety operators are employed by the industry organization, although their expenses are subsidized by the government. Passenger safety is thus entrusted to the shipping business, whose priority probably lies elsewhere.

It took the tragedy of the Sewol for everyone to realize that the industry's self-regulation was a formula for accidents. Before the Sewol set sail on April 15, its operators loaded it with 180 vehicles and 1157 tons of cargo, but grossly under-reported that it had only 150 vehicles and 657 tons of cargo. To evade inspectors' scrutiny, they removed water from the ballast so that the ship would float above the safety line. The overloading in combination with the ballast emptying made the ship prone to capsizing. The Sewol's regular captain, who had been replaced with a temporary hire for the voyage, testified in court that these were common practices and when he raised the issue with Chonghaejin officials, he was told that if he were to raise his voice, he should resign from his post. The ship's overloading and false report were exposed only after the Sewol sank, and it was on June 3 that the Gwangju District Court issued an arrest warrant for a senior vessel safety operator of the KSA's Incheon unit for negligence.

The collusion between the state and the Sewol's owner risked not only the passengers' safety but also the crew's. Most of the Sewol's crew were temporary contract workers, as is common practice among Korea's domestic maritime transporters. Lee Junsok, the Sewol's captain, for example, was a 69-year-old temporary hire contracted with a monthly salary of $2700 a little before the Sewol's departure. More than half of the crew, including the captain during the fatal voyage, were temporary workers with contracts of six months to a year. They were not only denied fringe benefits, but also adequate safety training. As if hiring temporary workers was not enough to trade passengers' safety for profits, Chonghaejin minimized its spending on crew training. It allocated a paltry $540 for the crew's safety education in 2013 whereas it spent $10,000 on entertainment and $230,000 on public relations (PR), clearly showing its priorities. And yet it is mostly this crew that stood trial for the deaths of the passengers.

The Sewol sank under the weight of the collusion between the neoliberal state that shed its responsibility to safeguard people's lives to private entities and the private entity that traded customers' safety for profits. The accident serves as a vivid reminder of what tragic consequences can result from government–business collusion. While collusion had existed under previous authoritarian regimes that sometimes sacrificed people's safety for profits, the Sewol incident reveals that the nature of the collusion shifted to give more power to business interests. The authoritarian developmental state shed some of its power as part of the IMF-imposed structural adjustment after the 1997 financial crisis. As the government transferred some of its power to plan, manage, and oversee the economy to private entities, its relative power gradually declined. By the time of the Sewol disaster, the government took a hands-off approach to overseeing such "private" entities as the KRS. Privatized entities with increasing boldness ignored government directives and warnings, and became more independent and aggressive in pushing their agenda.[4]

---

[4]Several years ago, for example, the KRS's chairman set up a subsidiary company a little before his retirement, and he moved to the subsidiary as its inaugural chairman after his retirement. The Ministry of Maritime Affairs sent a few tepid warnings, but they were ignored, something that could not have happened under the authoritarian developmental state.

Rescue Failures by the Disappearing State

One of the greatest mysteries surrounding the Sewol incident is that neither the crew nor the government, including the Coast Guard and the Navy, made serious efforts to rescue the passengers trapped inside the sinking ferry. Thirty-five passengers were picked up from the deck and airlifted by three Coast Guard helicopters. Most of the crew, including the captain, were rescued by the Coast Guard's Patrol Boat 123 that pulled up to the control room of the Sewol so they could jump to safety. The patrol boat returned later to rescue additional passengers. Most of the surviving passengers were saved because they jumped off the ship before it submerged and were pulled out of the water by fishing boats that happened to be nearby. Other than these, no one was rescued from the sinking ship by the Coast Guard or the Navy after about 10:25 am on April 16, 2014.

The next several hours, the "golden time" during which the passengers could have been saved, was notable for the absence of active rescue operations. The Navy's Ship Salvage Unit (SSU) and Underwater Demolition Teams (UDT), as well as the Coast Guard's special units, were dispatched but arrived late and did not engage in active rescue operations. Their failure was compounded by deadly instructions by the crew, which repeatedly broadcast instructions to the passengers to stay put and not leave the sinking ship, contrary to common sense. In another illogical instruction, they told the passengers to wear life jackets first and stay within their cabin. The instruction proved deadly when the ship capsized and the passengers were trapped underwater, for, wearing a personal floating device, they could not swim underwater to escape from their cabins. A majority of the passengers, high-school students, followed the crew's instructions to their detriment. Meanwhile, the crew, including the captain, were the first ones to abandon the ship, leaving behind over 300 passengers, who remained in their cabins as they were told.

The crew's failure was compounded by the Coast Guard. It dispatched Patrol Boat 123 to the Sewol, but although some crew members of the boat boarded the Sewol before it sank, they made no effort to rescue the remaining passengers or even to tell them to abandon ship. They limited themselves to rescuing the Sewol's crew. The captain of Patrol Boat 123 testified in court on August 13 that he "panicked so much that he forgot" to instruct his crew to move into the Sewol's cabins, adding that he was "so busy that he could not tell the passengers to evacuate the ship." Commissioner of the Coast Guard, Kim Sok-Kyun, did not do much better. Via the West Sea Coast Guard Task Force, he instructed Patrol Boat 123 to send its crew to the Sewol and "calm the passengers to prevent them from panicking." What is clear is that no order was issued from the top of the Coast Guard hierarchy to rescue the passengers before the ship sank. Video footage of the Sewol during the golden time shows Coast Guard boats circling around the slowly submerging ferry, effectively keeping away the fishing boats that had come to help save the passengers.

It was not just the fishing boats that were kept away. The Navy could not enter the scene of the accident to participate in the rescue operation for the first two days. In the meantime, Undine Marine Industries, an ocean engineering firm that specialized in offshore construction and marine salvage but that had no record of, or professional employees trained for, passenger rescue, emerged as the central rescue operator. The day after the accident, Undine was contracted by Chonghaejin at the recommendation of the Coast Guard, and took control of the rescue operations, sidelining not only volunteer rescuers but also the Coast Guard and the Navy. The problem is that it acted as if it were more interested in salvaging the ship than saving the passengers' lives. Indeed, its divers saved not a single passenger. Even when all the passengers remaining in the ship were presumed dead, it delayed retrieving the bodies of the dead for as long as 20 hours. The void left by the state was filled by a private company that found an opportunity to maximize its profits by extending its operations as long as possible.

The state as an organization whose fundamental mission is to protect people's lives and provide for their safety failed throughout the crisis.[5] Not only did it fail to establish an effective control center to mobilize national resources necessary for rescue operations, but it added to the chaos of the accident by creating obstacles to the rescue and spreading flawed information. Various units of the government created a total of ten headquarters in response to the Sewol's sinking, creating confusion as to the line of command and producing problems in communication among government units. Ministries of Security and Public Administration, Oceans and Fisheries, and Education set up their

[5] *Newstapa*, a trailblazer of Korea's investigative journalism, aired a program that exposes the absence of the state during the "golden hours." http://newstapa.com/news/201415446.

respective headquarters while the Coast Guard and Kyonggi Provincial Government also created theirs. During the critical initial hours, not only was the central coordination of rescue operations lacking but no credible information was available. Different entities reported different numbers of rescued passengers, and in what proved a fatal mistake, the Central Disaster Management Headquarters announced that 368 passengers were rescued at 1:19 pm, 4 hours after the ferry's sinking, when in fact over 300 of them were missing. It was only the day after the accident that all the government units involved agreed to establish the Pan-Government Accident Response Headquarters that unified the rescue operations and communication. By then, the "golden time" was over, and the remaining passengers were presumably dead.

The irony of ironies is that the potent power of the military deployed to save civilians was stopped by the weaker Coast Guard to create room for a private salvage company to operate. The Tongyoung, the state-of-the-art salvage ship recently acquired by the Navy, was ordered by Hwang Ki-Chol, Naval Chief of Operations, *twice* to sail to the scene of the accident and help the rescue operation, but it never left its port. Representative Kim Kwang-Jin suggested that "it was difficult to understand why the Tongyoung was not dispatched even if the naval chief of operations twice ordered emergency assistance." Kim Min-Sok, the Ministry of National Defense spokesperson, responded that it could not participate in the rescue operation because one of its critical components, a rescue submarine that was needed to retrieve passengers from the sunken Sewol, had not been sufficiently tested or certified for operation. Why then did the top naval commander make an uninformed order that the uncertified ship participate in the rescue? Who made the decision to effectively disobey the Naval Chief's order? Why was no one reprimanded for such a fatal snafu? All of these unanswered questions have led to speculation that a higher authority, one higher than the Naval Chief, played a role.

Many suspicious eyes turned toward the National Intelligence Service (NIS) for Nam Jae-Joon, its chief, was known to be one of the president's confidants and wielded more power than his official position might suggest. Two documents, retrieved from the Sewol, seemed to indicate that the NIS had been deeply involved in the operation and management of the Sewol. First, the Sewol's emergency contact diagram listed the NIS as the first point of contact in case of accident. It and its sister ship, the Ohamana, also owned by Chonghaejin, were the only ones among the 17 large passenger ferries that were required to report an accident to the NIS before any other agency such as the Coast Guard. Second, a laptop computer retrieved from the sunken Sewol had a document, titled "Items of Planned Work in the Ship's Passenger Area—List of Items Identified by the NIS," that listed approximately 100 items that the NIS ordered repaired.[6] Although the NIS later claimed that it ordered those repairs for security purposes, the list included such items as vending machine installation, recycling bin location, ceiling paint, ventilator clean-up, etc. to which a security agency would not normally pay attention unless it were involved in managing the ship. According to the list, the NIS even required the Sewol to submit an employee wage report and the crew's vacation plan as if it had been involved in operating the ferry. These documents fueled suspicions that the NIS, as the real owner of the Sewol, was implicated in a foul play (Fig. 1.1).

Why would the intelligence agency sacrifice a ship? Pointing to the fact that the NIS had been on a hot seat before the Sewol disaster, critics raised the suspicion that it needed a scapegoat to divert attention. The NIS had indeed been struggling to defend itself against mounting evidence that it had been deeply implicated in a fabricated spy incident. Yoo Woo-Sung, a Chinese Korean who had defected from North Korea to the South, was accused by the NIS of working as a North Korean agent after his defection. He responded by accusing the NIS of falsely charging him. It was discovered during a court trial that NIS officials had used pressure and inhumane treatment to force Mr. Yoo's sister to testify against him and some NIS officials had gone so far as to forge three Chinese official documents to present to the court as evidence. In response to a query by a defense lawyer, the Chinese Embassy relayed to the court an official statement that the documents were not authentic, creating a diplomatic fiasco. Added to the mounting evidence that the NIS had been involved in the last presidential election and other domestic politics, the spy fabrication case could become the last straw for the NIS leadership. Public outcry was so strong and grew so rapidly that on April 15 Nam Jae-Joon, Director of the NIS, had to make a public apology for the spy fabrication, although he also made clear his intent to stay on the job.

---

[6] 416가족협의회, "세월호 업무용 노트북 증거보전 관련 기자회견: 세월호 실소유자는 국정원?" 2014년 7월 25일, 목포지방법원 http://416family.org/1381; 조근영, "대책위" 국정원 세월호 운항·관리 깊숙이 개입"(종합)," 연합뉴스, 2014년 7월 25일 http://www.yonhapnews.co.kr/society/2014/07/25/0701000000AKR20140725176000054.HTML.

■ 별첨1. 국정원 지적사항

[선내 여객구역 작업예정 사항]-국정원 지적사항

2013년 2월 27일(수)

NO	작 업 내 용	작업자	비 고
1	갤러리룸(전시실) 천정칸막이 및 도색작업	거성종합	5-6일
2	자판기 로비층 테이블 설치 여부?	임차장님	보고
3	분리수거함 및 재떨이 위치선정	임차장님	보고
4	오락실 바닥 데코타일 신환 및 천정 도색작업	거성종합	수리신청서
5	예비휴게방 출입문 DOOR 상런지 및 유리창 보수	거성종합	수리신청서
6	레스토랑 유리 파손면 썬팅보수	거성종합	수리신청서
7	편의점 유리파손면 썬팅보수	거성종합	수리신청서
8	화장실 타일 및 변기 신환 공사	풍성산기	
9	여성샤워실 누수부분 용접 및 타일 마무리작업	풍성산기	
10	샤워실 배수구 분리작업	본선작업	
11	베드룸 입구 불량장판 보수작업	거성종합	우드타일
12	3층 선수 화장실 입구장판 보수작업	거성종합	우드타일
13	객실 내.외부 유리창 청소작업	청소용역	
14	여객구역 비상등 램프 교체작업	본선작업	
15	CCTV 추가 신설 수리신청(브릿지 LIFERAFT 2곳)	삼아ENG	수리신청서
16	CCTV 추가 신설 수리신청(트윈데크 2곳)	삼아ENG	수리신청서
17	객실내 일본어 표기 아크릴판 제거작업	본선작업	
18	탈출방향 화살표 제작 및 부착	정상기획	
19	커피숍 냉장고 FAN 불량 및 R-22 냉매보충 수리	더난터	수리신청서
20	커피숍 워터식 세척벨브 신환수리	더난터	수리신청서

◂ **Fig. 1.1** <Items of Planned Work in the Ship's Passenger Area—List of Items Identified by the National Intelligence Service (NIS)> A laptop computer retrieved from the sunken Sewol had a document, titled "Items of Planned Work in the Ship's Passenger Area—List of Items Identified by the NIS," that listed 100 items that the National Intelligence Service (NIS) ordered repaired. This document, together with the emergency manual, fueled suspicions that the NIS might have been the real owner of the Sewol. (*Source* 416 Family Association)

The day after his press conference, the Sewol sank, effectively diverting attention from the NIS scandal. But the diversion did not last long as evidence of the NIS's implication began to surface. The Sewol's emergency manual and repair work list, unexpectedly retrieved from the sunken Sewol, fueled suspicions about the NIS's role, and the media and the opposition parties focused attention on this. As public suspicions grew, Nam Jae-Joon, the intelligence chief, tendered his resignation on May 22, and President Park swiftly accepted it. Because neither he nor she offered a compelling reason for his sudden resignation, the Lawyers Alliance for Democracy pointed out in a statement that there were grounds to suspect that the NIS had been implicated in the Sewol's sinking and the failed rescue operation. Nam's resignation nonetheless helped shield the ruling Saenuri Party from political liability just 10 days before a local election. It also protected the intelligence agency from parliamentary inspection, for the top intelligence officer who was in charge of the agency at the time of the accident was no longer available to report to the Parliament's special committee on the Sewol.[7]

Thus the parliamentary special committee called on Chief of Staff Kim Ki-Choon, who was commonly viewed as the real power in the presidential office, to testify on the Sewol. While evading most questions and doing his best to clear the Blue House of any responsibility for the bungled rescue, he nonetheless revealed an important fact in response to questions about the president's whereabouts during the golden hours. He testified that he and other officials reported to the president via written reports or telephone calls but had no face-to-face meetings until President Park showed up in the Central Disaster Management Headquarters

[7] Nam's resignation did not stop suspicions about the role of the NIS. During a court trial on August 22, crew member Ha testified that an NIS agent was aboard the Sewol three days after its launching on March 15 last year. When asked if the agent pointed out such problems as overloading or loose ties, Ha replied he did not remember.

around 5 pm. Her appearance there after seven hours missing in action was nationally televised. So was her ignorant question: "If the students are wearing a life vest, why is it so hard to find them?"[8] Apparently she was unaware that they were trapped inside the overturned and submerged ship and thus could not be seen in the open sea. The President's daily log, later released via Representative Cho Won-Jin of the ruling party to quell questions about her whereabouts, only confirmed her absence, for it failed to list a single face-to-face meeting. What was she doing for the seven hours? Where was she?[9]

Wherever President Park may have been on April 16, it is more than clear that the state, from top to bottom, was absent from rescue operations during the golden hours when the passengers could have been rescued. What looked like a strong national security state failed to save people's lives from the danger it had created with deregulation and privatization. The Korean state proved a failed one when it came to protecting citizens' lives, a fundamental *raison d'état*. It is also possible that the national security state failed in another sense, for in the aftermath of the tragedy, President Park succeeded at a summit meeting with President Obama in keeping the wartime operational control over the Korean military in the hands of the commander of the Combined Forces Command—a US general—effectively depriving the Korean state of the ability to control its own military during wartime. Moreover, she agreed in return to "enhanc[e] the interoperability of our missile defense systems," possibly opening the gate to the 2017 decision to deploy the THAAD (Terminal High Altitude Area Defense) missile defense system

[8] Cheongwadaetv, the Presidential Office's official YouTube channel, shows President Park asking the question at 1:15 of the video clip https://youtu.be/zLn_sp91A0s?t=1m15s.

[9] Her being missing in action led to unconfirmed rumors about her private life during the early hours of the accident. The largest circulating conservative paper *Chosun* ran an op-ed by Choi Bo-Shik, one of its editorial writers, that gossip about her private life had become a newsworthy issue openly discussed. When Korea bureau chief Kato Tatsuya of Japan's *Sankei* newspaper picked up the gossip to write "the 'fact' that President Park's unknown whereabouts for 7 h on the day of the accident emerged to shake the regime," the Blue House quickly retaliated by filing a defamation suit. Now the Sewol created another diplomatic row between Seoul and Tokyo, which grew to become an issue pertaining to the freedom of press.

in South Korea.[10] Given the Korean state's failures and successes, one might be tempted to say the state was so busy undermining the foundation of national security that it failed to protect its own citizens.

The Families Demand Truth and the Strong State Refuses

The Sewol tragedy resulted from the collusion of Korea's sea mafia, neoliberal deregulation run amok, the intelligence agency involved in shadowy activities, and a president possibly distracted by her private life. This would make a good movie, except that it cost the lives of 324 people, mostly high-school students, leaving their surviving parents and relatives still grieving and searching for the truth. One of them staged a hunger strike for 46 days demanding an independent and thorough investigation. The victims' families demanded that a special law be enacted to create an independent committee with subpoena and prosecutorial powers in order to find the causes of the death of their beloved ones. They believed that creating an independent committee would be critical to finding an answer to questions about the Sewol's sinking and the government's rescue failures, because all other methods proved ineffective.

The national parliament on May 29 created a special committee to investigate the Sewol accident, but the committee proved dysfunctional from the beginning. Its operation was stymied by repeated clashes between the two main political parties, the conservative Saenuri Party and the liberal Democratic Alliance for New Politics. Furthermore, the ministries and agencies called to report to the special committee dragged their feet and revealed little that was new. The Blue House made an effective investigation difficult by releasing only 13 of the 269 materials requested by liberal members of the committee two days before it was due to testify. The committee ended its work without even holding a hearing. The special committee failed to bring out any significant new facts, much less the truth, about the administration's failures during the incident.

---

[10] Blue House, "Press Conference with President Obama and President Park of the Republic of Korea," April 24, 2014. https://seoul.usembassy.gov/p_rok_042514f.html. The THAAD decision placed Seoul in a difficult position for Beijing, seeing this as a US move to undermine its strategic forces, strenuously objected with veiled warnings of retaliatory measures. Also it had the effect of accelerating the arms race on the peninsula to the point where both sides openly threatened a first strike. For more on the THAAD deployment and its ramifications, see 서재정, "사드와 한반도 군비경쟁의 질적 전환: '위협의 균형'을 무너뜨리고 선제공격으로?" 창작과비평 168 (2015년 여름), 414–440.

The Board of Audit and Inspection (BAI), Korea's counterpart to the US Government Accountability Office (except that it is part of the executive branch of the Korean government) conducted its own audit that seemed designed more to exonerate the Blue House than to examine its failures. After auditing the Blue House, it concluded that the presidential office was not responsible for the Sewol failure. The sole basis for its conclusion seemed to be a Blue House statement that "the Blue House is not the control tower of disaster management." It turned out that the BAI sent a few low-ranking officials to audit the Blue House, and they completed their work without even examining the reports that had been submitted to the president on the day of the accident.

Prosecutors and the police produced more questions than answers. Prosecutors were more aggressive in pursuing Yoo Byung-Un and his family for bearing the ultimate responsibility as the real owners of Chonghaejin than in investigating the government's failure to rescue the passengers. Even their investigative work on Mr. Yoo was shoddy. They staged a nationwide manhunt to arrest him, even to the point of organizing neighborhood watch meetings throughout the country for the first time since 1996, but missed him in a search of a house where he was hiding, only to identify a dead body as his 40 days after the body was discovered. Not to be outdone, the police sent the Sewol's captain, Lee, to the apartment of one of its officers' for the night after the Sewol's sinking, and provided him with a safe haven for almost a day. Prosecutors thus far indicted some of the crew and lower ranking officials for various charges related to the Sewol's sinking, but turned a blind eye to the rescue failure.[11]

Their failure to bring out the truth was accompanied by efforts by the NIS and the police to silence the victims' families and their supporters. The police had monitored the victims' families when they held meetings and blocked them when they tried to march to the Blue House to make a direct appeal to the president. Riot police isolated the families and their supporters by surrounding them with a wall of police buses. An unidentified person reportedly snooped around the hometown of a victim's father in what looked like a fishing expedition. An NIS agent paid a visit to the hospital that employed a doctor who was helping the

[11] Kim Hansik, Chief Executive Officer (CEO) of Chonghaejin, was sentenced by the Supreme Court to seven years' imprisonment and 2 million Won in fine for negligent homicide and violation of the ship safety law. 김계연, "'세월호 책임' 청해진해운 김한식 대표 징역 7년 확정," 연합뉴스, 2015년 10월 29일.

victims' families, and met with its director to inquire about the doctor's background. As SNS and media spread negative rumors about the families, Representative Min Byung-Du raised the suspicion that "the rumors are being spread through specific channels created by an expansion and reorganization of what looks like the ruling group's psychological warfare unit that operated during the last presidential election campaign."

The Sewol Families Committee sent a letter to President Park on August 22, 2014. In it, the families pointed out that "there is a larger issue at stake than specific issues related to a special law" and "that is whether the truth will be revealed or hidden." "We have come to know that the Blue House stands at the center of the efforts to hide the truth," the families wrote. "The president said that the truth must be unearthed lest the families should have any remorse, but has even refused to submit materials to the audit by the parliament." The victims' families demanded the creation of an independent committee with subpoena and prosecutorial powers in order to find the causes of the death of their loved ones. Kim Young-O, father of one of the victims, even staged a hunger strike for 46 days to demand just that.

The Sewol families' demand for the truth, along with millions of citizens who supported them by signing their petition, bore fruit, albeit one that was deformed by pressures from the ruling party and further stunted by manipulations by the administration.[12] On November 19, 2014, the Special Law to Investigate the Truths of April 16 Sewol Ferry Incident and to Build Safe Society was enacted that would create the Special Investigation Commission on 4/16 Sewol Ferry Disaster to investigate the disaster, propose reform measures on safety, and provide support to victims, survivors, and their families.[13] While the commission was thus launched, it was made toothless from the beginning. Despite the families and many scholars' demands, it was deprived of subpoena and prosecutorial powers and thus forced by design to depend for its investigation on the goodwill and willing cooperation of officials and individuals, many of whom were suspected of wrongdoing. Limited as its power

---

[12] By July 8, 2014, 3.5 million citizens had signed the petition to enact a special law. 양영석, "세월호 국민대책위 9일 '세월호 특별법' 입법청원'" 연합뉴스, 2016년 7월 8일. By November 14, 2014, 6 million citizens had joined the petition. 이희훈, "세월호 유족 '특별법으로 끝? 안전사회 건설 이제 시작'" 오마이뉴스, 2014년 11월 14일.

[13] http://www.law.go.kr/unSc.do?menuId=10&query=%EC%84%B8%EC%9B%94%ED%98%B8%20%ED%8A%B9%EB%B3%84%EB%B2%95 Accessed August 17, 2017.

was, its ability to function as an independent entity was further compromised by the enforcement ordinance—the executive order to implement the special act was issued after a long delay, six months after the act—that empowered government officials, seconded from various ministries including the Ministry of Oceans and Fisheries and the Ministry of Public Safety and Security that were implicated in the disaster, to play a central role in managing the commission's operation and investigation.[14]

After it was launched with these limitations, its activities were then crippled by obfuscations from inside and out. One staff member seconded from a ministry, for example, leaked inside information to his ministry, the Blue House, the ruling party, and the police.[15] When the commission decided on November 23, 2015 to examine the appropriateness of the Blue House's response as part of its investigation, Saenuri Party members of the National Diet's Oceans and Fisheries Committee demanded that the commission members resign, and the commission members, chosen by the ruling party, did exactly that to protest the commission's decision to investigate the president. These responses seem to have been coordinated with the executive branch, as they, and more, had in fact been spelled out in a strategy paper that the Ministry of Oceans and Fisheries had prepared to lay out a series of effective measures to protect the Blue House from the investigation.[16] As if these limits and obstructions had not been enough, government ministries started actions by May 2016 to prematurely and permanently disband the commission. The Ministries of Interior, Planning and Finance, and Oceans and Fisheries sent the commission a notice that it should now wrap up the investigation and begin to prepare a final report because according to their interpretation of the Special Act, its term would expire by the

[14]The Enforcement Ordinance was criticized by many, including members of the ruling party such as Yoo Seungmin, for violating the special act that the ordinance was supposed to implement. Representative Yoo in response took steps to introduce an act that would prevent the executive branch from using an enforcement ordinance to undermine or violate the lawmaking power of the National Assembly, but was censured by President Park for "betrayal" so vehemently that he had to yield his position of Majority Leader and subsequently failed to be nominated by his own party to run for the next general election.

[15]황철환, "'세월호 특조위 내부자료, 당·정·청·경찰에 유출'(종합)" 연합뉴스, 2015년 3월 23일.

[16]박다혜, "[단독]해수부 "세월호 특조위, BH 조사시 與위원 사퇴 표명"...'대응방안' 문건," 머니투데이 뉴스, 2015년 11월 19일. http://m.mt.co.kr/renew/view.html?no=2015111908347632431#imadnews.

end of June. The act stipulated that the commission should complete its work within "one year after its formation" and could, if necessary, extend the period once by six month. The administration argued that since the act went into effect on January 1, 2015, the commission's term began then and could operate only until the end of June 2016. But the commission pointed out that it was formed on August 4, 2015 when its members were hired and its budgets allocated, and thus could operate until February 3, 2017. To protest what they viewed as the administration's willful violation of the special act and to continue what the law empowered them to do, Commission members started a relay hunger strike on July 27, and Sewol family members and other citizens joined them.[17] Their desperate efforts notwithstanding, the commission was disbanded on September 30, 2016 even before it could examine the body of the sunken Sewol.[18]

The Sewol families, the Special Commission, and millions of citizens have thus continued their efforts to remember the dead and find the causes of their death, proving Milan Kundera's often quoted maxim, "The struggle of man against power is the struggle of memory against forgetting."[19] The Park administration and the ruling Saenuri Party have responded by forcefully denying their demand for truth. What are they afraid of?

II. Compressed Modernization, Compounded Problems[20]

The Sewol sinking, a maritime accident, should not have become a societal crisis. That is at least in theory. It, however, evolved into such, exposing latent, yet explosive, contradictions that lay contained under the pressure of normalcy. A triggering crisis often leads to a clash between the banality of an ancient regime and the challenges of the new. The shifting boundary reveals a multitude of forces that maintain a tight lid over society, which remain mostly unseen under the presumed sense

---

[17] 금보령, "세월호 특조위 단식 39일차.. '지치고 힘들 땐 내게 기대'" 아시아경제, 2016년 9월 4일.

[18] At the time of writing, more than two years after the sinking, the wrecked body of the Sewol had not yet been salvaged due to a number of delays.

[19] Milan Kundera, *The Book of Laughter and Forgetting* (New York: Penguin Books, 1980), p. 3 (Kundera 1980).

[20] I gratefully acknowledge the contribution that Mikyoung Kim, as co-editor of the volume, made to developing this section.

of order. We define crisis as a state of comprehensive social dystrophy resulting from actualized risk potentials.

Crisis is both a result of past practices and an impetus for future transformation. It may appear instantaneous, but stands on the temporal continuum of the past, the present, and the future. It reveals the structure, hitherto unseen, and the past that shape the present. It also unleashes forces channeled by the structure of the present and yet capable of reshaping, undermining, or demolishing the establishment. Thus starts a contest between the incumbent and challengers, the outcome of which is yet to be determined. Crisis, in this sense, merely opens up an arena where the accumulated banality is challenged by the present disorder out of which the future is born. In its progression, some forces dissipate while others are born anew and yet others transformed. The arena created by crisis becomes a stage where the multitude of elements such as temporal spans (e.g., past, present, and future), powers, and norms engage in simultaneous contestations.

April 16, 2014 was such a crisis for Korea. That fateful day, the Sewol sank under the sea, bringing to the surface a multitude of forces that were active and yet remained unseen in the workings of the Republic of Korea. What might have been a traffic accident in the sea turned out to be much larger than it appeared at first—not only in terms of the number of deaths, the days of rescue operations, or the amount of lost assets but also in terms of the magnitude of the country's hazards that it exposed. It brought to the fore, in a way that cannot be ignored or dismissed, many hazards that the country itself has generated and yet ignored through the process of modernization. It has exposed shortcomings in the institutions, laws, practices, and relationships that had sustained the Republic for over five decades while at the same time crystallizing people's amorphous and unarticulated anxieties into a tangible, palpable disaster. The Sewol incident exposed the unseen past, articulated the unarticulated present, and shaped the future yet to come. It was such a crisis.

The nationwide crisis caused by the Sewol incident poignantly exposed tensions that have accumulated under Korea's accelerated development and quick liberal transition. Celebrations of South Korea's economic and political achievements notwithstanding, those successes are characterized by two seemingly simultaneous and yet contradictory processes: compressed modernization and compressed contradictions.[21] Not only has the former generated the latter, but also exacerbated them by keeping them unarticulated under the ever growing pressure to accelerate the modernization project of liberal economic and political development. While the process may have led to the wonders of newly—and quickly—acquired economic and political liberty that people savor, it has also placed their lives under growing risks resulting from its own success. There were ominous signs already before the Sewol incident that the risks were growing to an unsustainable level. The Seongsu Bridge fell down with the loss of 32 lives in 1994, and the Sampung Department Store in Seoul collapsed the following year, killing 502. In 2003, a subway train in Taegu caught fire and left 192 dead and 151 injured. In the seas, the Seohae Ferry sank in bad weather in 1993, resulting in the death of 292 passengers out of 362. Over two decades earlier, the Namyoung went under together with 326 passengers. Each accident, costing many a dear life, was a man-made tragedy resulting from such structural problems as institutionalized corruption, unbridled market power, growing and deepening stratification, and a debilitated civil society among others. Yet all were papered over, paving the road to the latest tragedy, the Sewol.[22]

What is troubling is not so much that these accidents occurred—accidents do occur everywhere after all—as that they failed to elicit serious self-reflexivity. This is not to deny that they led to strong outcries by the public and many remedial measures by the government. The outcries did not last or, more importantly, develop into a serious and sustained examination of the causes that might have resulted from the very modernization enjoyed by the public. The remedies included many stop-gap measures such as penalizing low-ranking officials as a scapegoat, and some of them were counterproductive as they exacerbated the

[21] Kyung-Sup Chang, "Compressed Modernity and Its Discontents: South Korean Society in Transition," Economy and Society, Vol. 28, No. 1 (1999), pp. 30–55. Kyung-Sup Chang, "Compressed Modernity in Perspective: South Korean Instances and Beyond," 5th World Congress of Korean Studies (Chinese Culture University, Taipei, Taiwan: 2010) (Chang 1999, 2010).

[22] For the Namyoung incident, its captain was sentenced to imprisonment for two and half years and its owner for six months and a fine of 30,000 Won. Only four low-ranking officials and Coast Guard officers were indicted but found innocent by the court.

underlying conditions. While after the Namyoung tragedy, 30 university students issued a self-reflexive statement on December 30, 1970 that the incident was "not a simple accident but resulted from the trend that belittles human life in the process of modernization," they were small in number and smaller in influence. President Park Chong-Hee turned a blind eye to their warning, as the rest of society did. At a cabinet meeting a week later, he instead railroaded his modernization plan headlong: "The Namyoung incident was caused by the lax discipline of the relevant officials. While officials' corruption is bad, worse is the laxity of their discipline. Tighten the discipline lest there should be any trouble in preparing the third five-year [economic development] plan.[23]"

Ulrich (1992) differentiates reflexive modernization from simple modernization propelled by industrialization. In the former, as a society becomes more modernized, members of the society reflect on the condition and effects of modernity and gain the capacity to transform it by themselves as reflexive civic subjects. The latter is characterized by the accumulation of industrialization measures implemented without critical examination of the consequences and without the formation of reflexive civic subjects. Korea comes closer to the latter than the former. Since the establishment of the republic in 1948, the government has pursued—and justified its choice in terms of the global Cold War that squarely placed it to the south of the divided peninsula—a capitalist economic growth without ever exercising self-reflexivity. The early experiment until 1960 had been marred by the rent-seeking endemic at the top of the sociopolitical hierarchy, who had all the incentives not to examine the potential ramifications of the capitalist trajectory. The military junta since 1961 took its economic development plans as seriously as it would conduct a military operation. It set specific targets, mobilized all the resources it could garner, and issued a marching order. It never paused, or allowed anyone to pause, to critically examine the premises or consequences of its economic growth plans lest a pause, or even a slowdown, should cause a criticism of not only its economic policy but also its own legitimacy. This is not to deny there were critics of the dictatorship and its policies even during the worst of the oppression—there were undoubtedly many—but to point out that even most of these critics shared with the regime a

[23] 김은남, "선장은 '3년형', 해경은 '무죄'... 남영호 판결," 시사인, 2014년 5월 21일. http://www.sisainlive.com/news/articleView.html?idxno=20308.

blind faith in modernization, differing only in methods, paths, and requisite politics. At least, they did not dare to deviate from the path forced upon their country by the global bifurcation and the peninsula division.

The "division system"—a concept developed by Nak-chung Paik to highlight, among others, domestic consequences of Korea's division—is useful here for exposing the structural constraints that facilitated the country's compressed modernization and at the same time impeded the development of reflexive civic subjects.[24] While the series of authoritarian governments at least until 1987 stifled the political freedom to criticize, division imposed a structural constraint on Korea's discursive space itself to leave critical examinations of the capitalist path itself outside the realm of intelligible debate. The division system, in other words, put South Korea on the path toward not just a simple modernization that failed to reflect on the condition of modernity but also a compressed modernization that precluded the condition of reflexivity.

While Chang discusses "compressed modernity" as a broad social concept, our discussion of Korea's compressed modernization has more overlap with what some scholars call "compressed development."[25] As late developers Germany and Japan achieved in compressed time scales what it took Britain centuries to accomplish, so did late developers like Korea seek to emulate their predecessors' success in an even more compressed time scale.[26] Such compressed development of Korea created the mirage of quick accumulation of wealth as a national preoccupation that ironically diverted the national gaze away from the ontological

[24] Paik, Nak-chung. *The Division System in Crisis: Essays on Contemporary Korea* (Berkeley, Los Angeles, London: University of California Press, 2011) (Paik 2011).

[25] Chang, Kyung-Sup. "Compressed Modernity and Its Discontents: South Korean Society in Transition." *Economy and Society* 28, no. 1 (1999): 30–55; Chang, Kyung-Sup. "Compressed Modernity in Perspective: South Korean Instances and Beyond." In 5th World Congress of Korean Studies. Chinese Culture University, Taipei, Taiwan, (2010). Whittaker, D. Hugh, Tianbiao Zhu, Timothy J. Sturgeon, Mon Han Tsai, and Toshie Okita. "Compressed Development in East Asia." ITEC Working Paper Series (2007) (Chang 1999, 2010; Whittaker et al. 2007).

[26] While Hirschman called the pattern of development in Latin America "late, late industrialization" to distinguish it from the experience of late developers like Germany, we use the expression "late developers" to refer to Korea and other Asian "tigers" that emulated a late developer Japan at a later time. Albert O. Hirschman, A Bias for Hope: *Essays on Development and Latin America*. New Haven: Yale University Press, (1971) (Hirschman 1971).

insecurity of the people. In the collective *ppalli ppalli* [fast forward] *modus operandi* that was forged under the national development frenzy, looking back was regarded as an impractical, therefore pointless, practice.[27] As developmental stages were collapsed in late developers, furthermore, Korea experienced industrialization and de-industrialization almost simultaneously as well as integration with global value chains, which complicated any effort to look back. For late developers, "development in the context of deepening global integration introduces levels of simultaneity and international interdependency that are quantitatively and qualitatively different from those of archetypical late developers.[28]" Not only does compression challenge the developmental state's ability to adapt to the new, but also confounds social capacity to comprehend and critically engage the present. Compressed modernization compounds old ills with new complications.

Rather than confronting these atavistic and modern illnesses expanded and exacerbated under Korea's compressed modernization, the country since the financial crisis of 1997–98 accelerated the very processes that had produced them. If neoliberalism was "creative destruction," as Harvey argues, its application to Korea was a creative disaster.[29] Not only was it prescribed by the International Monetary Fund, but the then democratized state embraced it with heart and executed it, ironic as it may sound, with all its interventionist tools. The state shattered pre-existing institutions of safety with deregulation and privatization. Although it strengthened its ability to supervise the financial sector that was directly responsible for the financial crisis, it removed or relaxed safety regulations that it had maintained in most other economic sectors. It privatized not only state-owned enterprises such as Korea Telecom but also allowed the use of contractors and subcontractors, who relied heavily on irregular workers, for many functions played by state agencies or public corporations. The neglect of work safety and social responsibility, which

---

[27] Cho Han, Hae Joang. "'You Are Entrapped in an Imaginary Well': The Formation of Subjectivity within Compressed Development, a Feminist Critique of Modernity and Korean Culture." Inter Asia Cultural Studies 1, no. 1 (2000): 49–69 (Cho Han 2000).

[28] Whittaker, D. Hugh, Tianbiao Zhu, Timothy J. Sturgeon, Mon Han Tsai, and Toshie Okita. "Compressed Development." Studies in Comparative International Development, no. 4 (2010), 461 (Whittaket et al. 2010).

[29] Harvey, D. (2007). Neoliberalism as creative destruction. *The Annals of the American Academy of Political and Social Science*, 610(1), 21–44 (Harvey 2007).

had been the norm under compressed modernization, was now deepened under the new legitimation of efficiency enhancement and profit maximization. The maladies which had grown under compressed modernization thus accelerated under neoliberalization. They were only kept unseen under the pressure of cut-throat competition within Korean society and without until the Sewol incident blew open the façade of a developed liberal society and exposed the deformities of Korea's compressed modernization.

Negative consequences of neoliberalism on society are well documented by now. Studies have shown that deregulation may help lower consumer rates and raise business profits but these "benefits" are achieved at the cost of the workers who work longer hours, make less money, and lose many non-wage benefits, as well as at the cost of a society that suffers directly from increased accidents and bears the burden of externalities.[30] Also privatization, another aspect of neoliberalization, has been shown to degrade safety and shift externalities onto society.[31] Exacerbating the safety consequences of deregulation and privatization is the drive to cost-cutting, the effect of which on safety is found "a recurrent cyclical phenomenon.[32]" The combined effect of deregulation and privatization is shown to decrease transparency, deflect corporate or state responsibility, and ultimately undermine safety governance in not only job security, and work health and safety, but also such diverse areas as transportation, food, and public health.[33] Recent cases of public safety

---

[30] See, for example on effects of deregulation on trucking, Michael H. Belzer, *Sweatshops on Wheels: Winners and Losers in Trucking Deregulation* (Oxford: Oxford University Press, 2000) (Belzer 2000).

[31] Peter Swan, "Neoliberal Rail Policies and Their Impacts on Public Safety," a paper presented at Neoliberalism's Threat to Safety and Transportation Workers' Response, Seoul, Korea, October 28, 2015 (Swan 2015).

[32] While the finding was made in a study on worker safety in the offshore oil industry, its insights go beyond the industry. Charles Woolfson and Matthias Beck, "The British Offshore Oil Industry after Piper Alpha," New Solutions (August 2000) 10: 44. See also Vernon Morgensen, ed., *Worker Safety under Siege: Labor, Capital, and the Politics of Workplace Safety in a Deregulated World* (New York: Routledge,) (Woolfson and Beck 2000; Morgensen 2006).

[33] Busch, L. (2011). The private governance of food: equitable exchange or bizarre bazaar? *Agriculture and Human Values*, 28(3), 345–352. Henson, S., & Humphrey, J. (2009). *The impacts of private food safety standards on the food chain and on public standard-setting processes.* Rome: FAO/WHO (Busch 2011; Henson and Humphrey 2009).

breakdown in Korea—deaths and illnesses caused by such disparate factors as MERS break-out, Oxy RB humidifier sanitizers, microparticle pollution, and subway accidents—are not isolated incidents but have a common root in compressed modernization and its attendant compounded maladies. They must be seen as a warning that cumulated maladies can no longer be cordoned off.

This volume is divided into the three sections that respectively address the themes of society, governance, and subjectivities. These sections revolve around one common concern: what does the Sewol incident tell us about Korea's compressed modernization and compounded crisis.[34] While the country arguably made a decisive turn to a liberal democracy in the late 1980s and a neoliberal economy a decade later, the two achievements, a culmination of Korea's compressed modernization, have generated countercurrents while at the same time being pulled back by reverse currents. The deregulation and privatization, imposed on Korea under duress of the 1997–98 financial crisis and implemented with renewed vigor in recent years, have magnified the tendency to threaten the safety and lives of people. Not only has the liberalization of Korea's governance facilitated the neoliberalization of the economy, but it has also contained within itself old illiberal practices that worsen the negative impacts. Finally, it shows how individuals participate in identity constitution processes, out of which some reinforce neoliberal identities and others evince the possibility of alternative subjectivities. The turbulence resulting from the compounded crises, to speak figuratively, sank the Sewol, serving as both a manifestation of the accumulated pressures and a warning of a future disaster that, if left uncorrected, could hit the Republic of Korea anytime.

The first section addresses social consequences of neoliberal restructuring in Korea. Lee makes a poignant argument that the tragedies of the Sewol were not a mere accident or aberration, but indeed an unavoidable consequence of neoliberalism. While deregulation and privatization may result in an improvement of economic efficiency for businesses, they debilitate social safety regimes by removing the very means of monitoring and regulating profit-driven business practices. As profits are generated at the expense of social safety, argues Kang, the society as a whole gets addicted to the neoliberal mode of operation. Social institutions and

[34] The latest MERS (Middle East Respiratory Syndrome) outbreak in South Korea also reveals the same kind of social and political ills.

individuals take the irresponsible state for granted at the cost of their own welfare and safety. They themselves imitate the powerful and get addicted to the new norm where no one takes responsibility or adopts corrective measures even as they are getting consumed by the self-endangering addictive behavior. This perpetuates the precarious cycle of cutthroat competition, abuse of power, and irresponsible toleration and replication of the abuse. The pervasiveness of an abusive social relationship—the so-called *kap-eul kwankye*—is an outcome and a symptom.[35]

The second section analyzes the roles of the state implicated in the neoliberalization process. Neoliberalization is not a natural, or inevitable, process by which markets expands their power and reach, contrary to what many theories of neoliberalism suggest. Nam demonstrates that economic neoliberalization does not necessarily facilitate further political liberalization but has rather created room where the authoritarian features of Korea's democracy can rear their ugly head. You and Park develop Nam's argument further in Chap. 7 to show that the reversal of democratic consolidation has a longer historical root. While the Sewol sinking can be immediately traced to the corrupt and collusive actions by the ship operators and the overseeing authorities, as Yoonkyung Lee shows in her chapter, they note that these actions betray the institutional co-optation established by the earlier authoritarian regime. The resilience of the authoritarian institution, exposed in the Sewol incident, is further buttressed by state paternalism and hierarchical culture, according to Park in Chap. 8. He ties the increased efforts to silence criticisms of higher authority during the Sewol crisis to the pervasive practice to control the freedom of expression in South Korean society.

The third, and last, section examines how individuals have constructed different subjectivities in their interactions with the economic and political neoliberalization discussed in the first two sections. From the fact that high-school students made up the majority of the victims in the Sewol sinking, Moon links the superordinate emphasis on academic performance to the disciplinary mandates of capitalist logic. Primarily the youths, but larger society as well, have been domiciled to follow uncritically the orders from the market and state power to function as efficient subjects. But neoliberalism's disciplinary power is not uniformly

[35] Originally a legal term that refers to a contractual relationship between two parties, the expression is now used to highlight the abusive nature of a relationship where the party with more power maltreats the other.

exercised, argues Lee in Chap. 10. Different categories of citizenship are created by the state to differentiate subjectivities who are entitled to differential services and protections. A majority who is disciplined to become neoliberal subjects, as described by Moon, therefore, is placed in a higher stratum of Korean citizenship, enabling it to feel safer than a minority in a lower classification, Hyunok Lee shows. A Vietnamese-Korean citizen, for instance, remains invisible and yet is critical to the reproduction of Korean citizenry who embrace neoliberalization with heart.

Last, but certainly not least, Hyeon Jung Lee shows one of the unintended consequences of Korea's neoliberal restructuring. The bereaved families of the Sewol victims have, through the process of collective grieving, transformed themselves from passive citizens who had mindlessly accepted the state's policies to active subjects who challenge and resist it. In their search for an answer as to why they had to lose their loved ones in the Sewol sinking, they have risen above the isolated comfort zone of the neoliberal individual subjectivity to connect with others alienated, sidelined, or dismissed by deregulation and privatization.[36] They hint at the possibility that neoliberalism may give birth to a new subjectivity that serves as a conduit to an alternative future. This possibility suggests healthy dialectics where the authoritarian and neoliberal past is confronted by emerging forces of a new catalyst.

Korea, seen through the prism of the Sewol tragedy, manifests compressed modernization in the raw. It has gone through an accelerated economic growth, transforming itself from a semi-feudal agrarian society to a neoliberal industrial/postindustrial one within less than a century. It has moved from a hereditary monarchy to an authoritarian and to a liberal democratic entity, still containing the residuals of the bygone era under the modern governing system. The neoliberal transformation under Korea's liberal governing system has been compressed as well; first, because Korea's structural adjustments during 1997–1998 were mandated by the International Monetary Fund, and second, because the Korean economy was caught between the narrow and narrowing space

---

[36] It is instructive to note a parallel with how miners in Wyoming strengthened their ties and affirmed their collective responsibility as they countered the individualization of risk promoted by neoliberalism. Jessica Smith Rolston, "Risky Business: Neoliberalism and Workplace Safety in Wyoming Coal Mines," *Human Organization* (2010) 69 (4): 331–342 (Rolston 2010).

between the graying advanced economies of the West and the rising economies of Asia. New problems created by such complicated transformations, and tensions between the emergent and old problems, have been numerous and yet compressed under the pressures of global and domestic competition. The Sewol incident was the crisis trigger that released the compressed contradictions.

Korea, as such, is both *sui generis* and general. The antinomies exposed by the Sewol reflect the particulars of Korea's development. At the same time, they reveal the potential tensions inherent in other societies on a similar developmental trajectory, albeit at a slower pace and under less pressure over a longer time horizon. This volume on the Sewol tragedy is to reflect upon multilayered implications that the disaster reveals about potential human costs of the state-driven project to enable market forces to penetrate into society deeper and faster than otherwise. If Karl Polanyi observed a "great transformation" in mid-nineteenth–early twentieth century Europe where an expansion and deepening of the market was accompanied by political intervention to safeguard society, we might be observing a "compressed transformation" in the late twentieth–early twenty-first century Asia where an expansion and deepening of the market is facilitated by political intervention to the cost of society.[37] It is perhaps no accident that the East Asian region in particular has experienced an unprecedented kind and level of disasters such as the 2008 Sichuan earthquake in China and the 2011 Tohoku earthquake–tsunami–radiation fallout in Japan. A very similar boat sinking accident took place in China on June 1, 2015, claiming more than 400 lives. These disasters differ in their origins, and yet share the commonality of a high social cost resulting from a compressed transformation. This volume on South Korea's Sewol incident serves as a helpful case study for the growing interest on the sanguine topics of disaster, governance, economic system, civil society, meaning-seeking and meaning-giving activities, ideational transformation of human agents and empowerment, and civil rights issues among others.

Korea has reached SOMEWHERE after having walked long and fast enough, as the Cat told Alice in *Alice's Adventures in Wonderland*. The question is where it has reached. It is the question that Koreans have

---

[37] Karl Polanyi, *The Great Transformation: The Political and Economic Origins of Our Time* (Boston, MA: Beacon Press, 2001) (Polanyi 2001).

been too preoccupied with compressed economic growth to pause to ponder. And it now confronts them. The Sewol tragedy reminds them the question may only be ignored any longer at their peril. We have thus attempted to seriously address it in this volume under a sense of urgency aptly expressed by Primo Levi: "It happened, therefore it can happen again: this is the core of what we have to say."[38] We hope our volume helps renew the sense of urgency among the public and contributes to the process of critical self-reflection.

References

Belzer, M. H. (2000). *Sweatshops on wheels: Winners and losers in trucking deregulation*. Oxford: Oxford University Press.

Busch, L. (2011). The private governance of food: Equitable exchange or bizarre bazaar? *Agriculture and Human Values, 28*(3), 345–352.

Chang, K. (1999). Compressed modernity and its discontents: South Korean society in transition. *Economy and Society, 28*(1), 30–55.

Chang, K. (2010). Compressed modernity in perspective: South Korean instances and beyond. In *5th world congress of Korean Studies*. Taiwan: Chinese Culture University.

Cho Han, H. J. (2000). You are entrapped in an imaginary well: The formation of subjectivity within compressed development-a feminist critique of modernity and Korean culture. *Inter-Asia Cultural Studies, 1*(1), 49–69.

Giddens, A. (1999). Risk and responsibility. *The Modern Law Review, 62*(1), 1–10.

Harvey, D. (2007). Neoliberalism as creative destruction. *The Annals of the American Academy of Political and Social Science, 610*(1), 21–44.

Henson, S., & Humphrey, J. (2009). *The impacts of private food safety standards on the food chain and on public standard-setting processes*. Rome: FAO/WHO.

Hirschman, A. O. (1971). *A bias for hope: Essays on development and latin America*. New Haven: Yale University Press.

Kundera, M. (1980). *The book of laughter and forgetting*. New York: Penguin Books.

Levi, P. (1988). *The drowned and the saved, Translated by Raymond Rosenthal*. New York: Summit Books.

Morgensen, V. (2006). *Worker safety under Siege: Labor, capital, and the politics of workplace safety in a deregulated world*. New York: Routledge.

[38] Primo Levi, *The Drowned and the Saved*, Translated by Raymond Rosenthal (New York: Summit Books, 1988), 199 (Levi 1988).

Paik, N.-C. (2011). *The division system in Crisis: Essays on contemporary Korea.* London: University of California Press.

Polanyi, K. (2001). *The great transformation: The political and economic origins of our time.* Boston, MA: Beacon Press.

Rolston, J. S. (2010). Risky business: Neoliberalism and workplace safety in wyoming coal mines. *Human Organization, 69*(4), 331–342.

Suh, J.-J. (2014). The failure of the South Korean national security state: The Sewol tragedy in the age of neoliberalism. *The Asia-Pacific Journal, 12*(40), 1.

Swan, P. (2015). Neoliberal rail policies and their impacts on public safety. In *A Paper Presented at Neoliberalism's Threat to Safety and Transportation Workers' Response*, Seoul, Korea, October 28, 2015.

Ulrich, B. (1992). *Risk society: Towards a new modernity.* London: Sage Publications.

Whittaker, D. H., Zhu, T., Sturgeon, T. J., Tsai, M. H., & Okita, T. (2007). *Compressed development in East Asia.* Institute for Technology, Enterprise, and Competitiveness (ITEC) working paper, 07–29.

Whittaker, D. H., Zhu, T., Sturgeon, T., Tsai, M. H., & Okita, T. (2010). Compressed development. *Studies in Comparative International Development, 45*(4), 439–467.

Woolfson, C., & Beck, M. (2000). The British offshore oil industry after Piper Alpha. *New Solutions, 10,* 44.

Author Biography

Jae-Jung Suh is Professor at International Christian University in Tokyo, Japan. He has served as Associate Professor and Director of Korea Studies at SAIS, Johns Hopkins University and Assistant Professor in the Department of Government at Cornell University, as well as on the Presidential Commission on Policy Planning (Republic of Korea). An expert on US–Korea relations, US policy toward Asia, international relations of East Asia, international security, and IR theory, he has authored and edited numerous journal articles and books, including Power, Interest and Identity in Military Alliances (2007); Rethinking Security in East Asia: Identity, Power and Efficiency (2004); Truth and Reconciliation in the Republic of Korea: Between the Present and Future of the Korean Wars (2012); and Origins of North Korea's Juche: Colonialism, War, and Development (2012), among others.

CHAPTER 2

The Sewol Disaster: Predictable Consequences of Neoliberal Deregulation

Yoonkyung Lee

When scholars and commentators discuss the consequences of neoliberal policy shifts, they focus mostly on deregulation in various economic spheres, privatization of public corporations, heightened economic instability at the macro level, and aggravated personal insecurity at the micro level. Living through the neoliberal era, we have learned a lot about how deregulation of all varieties has created crises in the financial market (e.g., 1997–1998 Asian financial crisis), the housing market (e.g., 2008 Lehman Brothers bankruptcy), and the labor market (e.g., the rise of precarious workers and economic disparities). Up until recently, however, the consequential relationship between neoliberal policy change and the occurrence of massive disasters remained elusive or hypothetical at best. This was especially the case for many Koreans until April 16, 2014 when the Sewol ferry suddenly capsized and took the life of 304 passengers, most of whom were teenage schoolkids on board for their field trip to Jeju Island. The Sewol accident and the way in which the crews, the ferry company, and state agencies responded to the accident

Y. Lee (✉)
University of Toronto, Toronto, Canada

J.-J. Suh and M. Kim (eds.), *Challenges of Modernization and Governance in South Korea*, DOI 10.1007/978-981-10-4023-8_2

provided an analytical magnifier to understand this not-yet-proven relationship between neoliberal deregulation and man-made massive disasters.

The Sewol disaster gave an unspeakable shock to Korean citizens and observers abroad. Perhaps the disaster's sociopolitical magnitude to the Korean public was comparable to that of Hurricane Katrina in American society. It was not the accident itself because accidents do occur in every society. The source of the shock was tied to how state authorities and responsible actors responded to the accident. Investigating the institutional conditions that led to massive disasters like Sewol or Katrina and chronicling the ways in which public authorities responded to and managed disasters reveal the strengths and weaknesses that constitute a society and its political systems. What was most appalling in the case of the Sewol was not the fact that a ferry capsized and sank but that an unending series of avoidable mistakes, misconduct, inertia, and incompetency was made by individuals, private interest, and state institutions. What surfaced minute after minute from the moment of the ferry's sinking were not saved passengers but the structural predicaments of Korean society and the comprehensive incapacity of public authorities.

On April 16, 2014, Sewol was declared seaworthy by the Inchon port authority despite the ship being full of built-in conditions for an accident. When the ferry suddenly capsized, the crew abandoned the ship and its passengers. No passenger from the ferry's cabin was rescued. The rescue mission by the Coast Guard and the emergency headquarters formed by the central government was a complete failure marred by incompetency, deception, and collusion. This study, instead of approaching the Sewol disaster as a random accident caused by the misconduct of a few individuals with failing morals, situates the Sewol in the larger political and economic context of Korean society. It argues that the Sewol accident reveals the disastrous consequences that the state's neoliberal deregulation and collusion with private interests have for public safety.

This chapter highlights two connections between the Korean government's embracement of neoliberal policies since the 1990s and the tragic accident of the Sewol. The first is that its neoliberal policies have created broad institutional flaws for public safety, thus increasing the likelihood of massive disasters like Sewol. The second is that they have undermined the state capacity to manage mega accidents when they occur. In investigating the neoliberal connections to the Sewol tragedy, this study chronicles the continued withdrawal of state functions from public safety regulation and oversight as evidenced by the deregulation of vessel management and safety assessment, and the privatization of rescue missions. It also identifies the social cost of the rising levels of labor market deregulation as exemplified by the expansion of temporary crew employees in the ferry industry and their inadequate response to disaster circumstances. Yet, in explicating the intended consequences of deregulation, this exposition also emphasizes how the old practices of state collusion with private business interests and the absence of transparent oversight thrived under the conservative government in Korea. Thus, the Sewol was caught in the combined deadly force of the seemingly new wave of neoliberal deregulation in recent decades and the continuity of public–private collusion originating from the developmental past. This chapter will conclude with some discussion on neoliberal deregulation, the role of public institutions, and the intrinsic tension between market interests and democratic justice.

Deregulation in the Ferry Industry

Korean society began to embrace neoliberal globalization and various forms of deregulation in the 1990s. The neoliberal shift started with the Kim Young-sam government (1993–1997) that pitched for Korea's *segehwa* (internationalization) and accelerated under President Kim Dae-jung (1998–2002), who administered the restructuring of the Korean economy in the aftermath of the Asian Financial Crisis in 1997–1998.[1] Former President Roh Moo-hyun (2003–2007) described his years in office as being besieged by the interest of capital to indicate the extent of market forces dominating political affairs.[2]

The neoliberal change further expanded with the return of conservative political forces to power. Without question, the government under Lee Myung-bak (2008–2012), a former CEO of Hyundai Construction, stood as an ardent champion of business-friendly policies. Under the banner of "MBnomics", the Lee government formed the Presidential

[1] For detailed discussion of economic changes during Korea's neoliberal era, please read Lee (2014).

[2] In his meeting with CEOs of Korean conglomerates in May 2005, President Roh made a comment that "the reigning power now lies in the hands of the market" (Hankyoreh shinmun, May 30, 2005).

Commission for National Competitiveness and introduced various relaxations in the areas of *chaebol* regulation (such as the relaxation of the total amount of cross-shareholding, the relaxation of the total investment amount ceiling system, and the relaxation of corporate ownership of financial entities and mass media) and tax rates (such as the reduction of corporate tax and income tax) (Lee 2012). Although President Park Geun-hye campaigned on "economic democracy" during the presidential race in 2012, her cabinet after coming to power completely erased the "democracy" part.[3] President Park continued the economic policy of her predecessor by subscribing to the pro-growth and pro-business emphasis. Deregulation continued under the Park administration as Ms President defined regulation as "our enemy and a cancer [for the economy]" when pronouncing a list of deregulatory measures (Redian, May 8, 2014).

The consequences of such a relentless process of neoliberal transformations were not limited to the restructuring of the Korean economy and the labor market, which championed business interests and produced an increasing number of precarious jobs. Neoliberal ideas that promote market competition and efficiency have deeply penetrated into political, social, and cultural institutions (Harvey 2007). A concomitant result of neoliberalism, therefore, was the creation of a social norm that justifies the prioritization of corporate profits and monetary gains over all other democratic values of human rights, justice, fairness, and transparency. It was in this context that state functions for the oversight of public safety were modified and the supposedly arm's-length relationship between public authorities and private interests were compromised.

The deregulation of vessel management and safety assessment, which this study identifies as the initial cause of the Sewol accident, needs to be understood within this broad context of the conservative government's drive for economic liberalization and deregulation. The KSA (or *Haewoonjohap* in Korean) is the organization that represents the interests of 1788 members who engage in the coastal shipping industry. The KSA began lobbying the Ministry of Oceans and Fisheries (MOF) in 2006 to extend the age limit of ships from 20 to 30 years.[4] In 2009, the Lee government passed the bill to extend passenger ships' lives to 30 years, along

[3]Economic democracy (*kyeongje minjuhwa* in Korean) denotes to economic redistribution and social welfare in the Korean political lexicon.

[4]The KSA website: http://www.haewoon.or.kr/ksaEng/main/main.do (accessed on December 17, 2014).

with various other deregulatory ordinances that would benefit coastal shippers (Kyunghyang shinmun, May 15, 2014). In addition to the extended ship's age, the frequency for engine inspection, specific guidelines for freight loading, overload limits, and the legal obligation of shippers involved in an accident were all relaxed over the years (ibid.). As the extension of vessel age went into effect, coastal shippers began to import used and aged ships (reaching about 20 years old) from other countries and to have them remodeled and renovated to put into operation in the Korean seas. The Sewol was one of the aged ferries, originally built in 1994 in Japan and put into retirement in 2012. The Cheonghaejin Marine Corporation bought the Sewol ferry in 2012 and refurbished it to accommodate a larger number of passengers and a greater weight of cargo in order to generate a greater profit.[5]

However, it is not to claim that the extension of ships' age per se is the primary cause of the Sewol accident. There are aged ships that sail in sound safety conditions. The safe operation of a vessel depends on the professional maintenance of the vessel and strict enforcement of vessel safety. Many countries, including Japan, have no age limit of vessels but instead have agencies that strictly enforce safety inspections and assessments (PPSS, May 14, 2014). Regulating vessels' age limit is an easy substitute for the state function for safety inspection and oversight. The problem with the Sewol is the combination of vessel age deregulation and lax enforcement of safety assessment by state agencies. This point leads to the second important cause that lies behind the Sewol disaster.

The KSA, the interest group that lobbied for the relaxation of state regulations on vessel age and operation protocols, is the association in charge of vessels' safety inspection. There are two additional organizations that are involved in the area of vessel safety. They are the KRS (or *Hankukseonkeup* in Korean) and the regional offices of Maritime Affairs and Port Administration (*Hangmancheong* in Korean) under the MOF. The KRS is a non-profit, private entity that specializes in ship classification and safety certification. The port authorities of Maritime Affairs and Port Administration are mainly in charge of port logistics, maritime traffic control, and inspection of cargo registration.[6] These three organizations, while the KSA and KRS are private organizations and port

[5] The remodeling of Sewol made the ship dangerously top-heavy.

[6] Information retrieved from the website of each organization: http://www.krs.co.kr and http://www.portincheon.go.kr (accessed December 17, 2014).

authorities are public ones, are in fact in a close relationship, mutually connected through personal ties and material interests.

First of all, when the Cheonghaejin Marine Corporation remodeled the Sewol and applied for safety certification, the KRS passed its safety assessment without raising a red flag. The remodeling raised the Sewol's center of gravity and required a reduced weight of cargo with an increased amount of ballast water. The KRS failed to make a note of such a structural problem. Second, it is basically the shippers themselves (the KSA and its members) who self-oversee their own affairs at ports and at sea. On the day of the accident, it was the cargo workers employed by Cheonghaejin who loaded the Sewol with twice the maximum allowed cargo. It was the chief mate who ordered the removal of several hundred tons of ballast water to conceal the overloaded cargo. Third, the Inchon Port Administration completely failed in its regulatory and oversight obligations. On April 16, the Port Administration officer signed off the Sewol's clearance notice even though the number of passengers and the cargo weight were left blank in the paper. The cargo worker employed by Cheonghaejin testified that no one has ever checked the actual weight of the cargo or corrected this violation of vessel safety regulation (Hankyoreh 21, March 25, 2016). He further revealed that the monthly inspections conducted by the KSA and the biannual special inspections jointly conducted by the Port Administration, the KRS, and the Korean Coast Guard (KCG) were merely perfunctory (ibid.).

This means that due to the KSA's close relationship to the KRS and the Port Administration, there is no active state agency involved in overseeing the safety inspection of ships, enforcing the compliance of safety regulations, and deterring the formation of collusive interactions. When no check mechanism is instituted between profit-seeking individuals or businesses and public safety, it is obvious that public safety will be compromised as it is perceived as an extra expense that reduces the maximum amount of profit for business owners. One of the fundamental causes leading to the Sewol disaster was the fact that neither a state agency nor a third party was actively present in overseeing the safety of vessel operation.

In addition to the problem of shippers' self-inspection of vessel safety, other measures related to safety continued to be deregulated under the Park government. The Ordinance on the Vessel Crew was revised to change the status of vessel inspectors and maintenance engineers from crew to non-crew members. This was because the existing Law on

Dispatch Workers prohibits using temporary dispatch workers for on-board crew members (Kyunghyang shinmun, May 15, 2014). By changing the status of vessel inspectors and maintenance engineers from crew to non-crew members, the shippers can employ temporary dispatch workers for this crucial function on the ship. As I will discuss in detail in the following section, the majority of the Sewol's officers and crew were temporary contract workers. These deregulatory measures implied that vessel safety was placed under double jeopardy. Not only are the shippers themselves in charge of vessel safety assessment, but they also employ temporary inspectors and engineers to be on board to respond to maintenance and safety problems at sea.

Labor Market Deregulation and Precarious Labor

The replacement of regular employees with temporary workers in the shipping industry is another predicament in public safety that this study identifies. The increased deployment of temporary workers in sea transportation is the small tip of an iceberg in Korea's labor market deregulation that has accelerated since 1996. The Kim Young-sam government pronounced the "New Labor Relations Plan" and formed the Labor Relations Reform Commission to increase the Korean economy's international competitiveness and to prepare the nation for membership of the Organization for Economic Cooperation and Development (OECD).[7] Capital, represented by the Korea Employers' Federation, pursued labor market deregulation from the very beginning of the tripartite negotiations (Choi et al. 2000). The revised labor law, passed in December 1996, overturned the compromise that was reached in the tripartite deliberations and instead introduced blatant clauses to make layoffs easy and expand the scope of contingent employment. The labor market restructuring continued under the Kim Dae-jung government that formed the Korea Tripartite Commission in 1998. The three parties (the government, capital, and labor) signed a social pact the same year but the essence of the pact was to introduce stipulations for massive layoffs and labor market flexibility in exchange for the legal recognition of the Korea Confederation of Trade Unions and labor unions' rights for political activity.

[7] The Commission consisted of three parties, i.e., the representatives of the government, capital, and labor, and deliberated on labor law reforms.

As a result of the labor market restructuring, the Korean labor force came to be composed of a massive number of non-regular workers (*bijeongkyujik* in Korean) (Lee 2014).[8] Many firms, even after the Korean economy's recovery from the financial crisis of 1997–1998, decreased or halted new hires of full-time employees with secure labor contracts. Firms substituted existing regular jobs with contingent workers contracted with precarious labor conditions. According to the definition of non-regular workers by a labor research group, the ratio of non-regular workers soared from 43.2% of all paid employees in 1996 to 58.4% in 2000 and has since been in a gradual decline to record 48.3% as of 2012.[9] What is phenomenal about non-regular workers is not only their size in the labor force, but also the gap they experience in their wages and social benefits. A non-regular worker, even if s/he performs the same labor as a regular worker, is paid 51.3% of a regular worker's hourly wage on average (Additional Survey of Economically-Active Population 2013). Furthermore, only about one third of over 6 million non-regular workers receive overtime payment, bonuses, and retirement payment, or are covered by paid holidays, national pension, national health insurance, and the national unemployment program. This is a huge contrast to regular workers as almost 90% of them are entitled to the aforementioned benefits and social protection (Lee 2014).

The ferry industry is not an exception but rather an example that demonstrates the extent to which businesses rely on contingent labor to cut labor cost. On the day of the accident, almost half of the crew on the Sewol ferry, including the captain, consisted of temporary contract workers, not full-time employees of Cheonghaejin, the ferry company. Out of a total of 29 crew members, 15 were temporary workers with short-term contracts ranging from 6 to 12 months (Pressian, April 21, 2014). Lee Joon-seok, the captain, was a one-year contract worker

---

[8] *Bijeongkyujik* includes long-term and short-term temporary workers, part-time workers, dispatch workers, subcontract workers, on-call (dispatch) workers, and others without a secure labor contract (Kim 2012).

[9] This study uses nonregular, precarious, and contingent labor/worker interchangeably. The government and labor activist groups have different definitions for nonregular workers. The former includes temporary, part-time, and atypical workers, whereas the latter adds long-term temporary workers and seasonal workers to the government classification. Data on nonregular workers come from Statistics Korea's *Survey on Economically Active Population* (various years).

who was paid about 60% of the pay of a full-time captain of the same rank (ibid.). Eight out of ten deck crews were temporary workers while three helmsmen were also contract workers. These temporary crews are also known to be paid about 60–70% of the wages for full-time crews (ibid.). These short-term ferry crews received no adequate safety training as Cheonghaejin spent just about $500 on safety education and training for the entire year of 2013. In comparison, the ferry company reported to have spent $60,000 on lobbying and $230,000 on advertising (ibid.).

These facts demonstrate that shipping companies were able to raise their profits by employing short-term contract workers and cutting costs for safety training. In a flexible labor market where hiring and firing is easy, employers view their workers as disposable goods instead of human resources requiring adequate training and long-term relationships. As the Cheonghaejin's meager spending on safety education shows, firms increasingly view workers worthy of only an hourly (or monthly) wage, while regarding any additional expenses spent on training or social protection as unnecessary overpayment. In industries where public safety is a key matter for daily operation, the provision of training, education, and other protective measures is an essential element. Such investment would enable workers to be professionally prepared with safety protocols and incentivize them to make a full commitment to job obligations. The captain and crew of the Sewol completely failed in following basic protocols in the crucial moments of the ship's capsizing and abandoned their professional obligations of saving the passengers in the cabin. While their misconduct is enraging and incomprehensible, blaming their failed morals only misses the broad context of the accident. Under precarious working conditions and in the absence of proper investment in training, it is unrealistic to expect workers to be fully committed to their work and to be professionally prepared for emergency situations.

The State's Withdrawal from Public Safety and the Ascendance of Private Interest

The state and public institutions have increasingly been placed under organizational disarray and thus incapacitated in public safety functions over decades of neoliberal deregulation. First, state agencies in charge of marine safety and regulation were placed under constant reorganization to serve the short-term political agenda of the incumbent president. The MOF was first established in 1996 along with the KCG as an affiliated agency. In 2008, President Lee introduced cabinet restructuring and dissolved the MOF by splitting it into the Ministry of Land and Oceans and the Ministry of Agriculture, Forestry, and Fisheries. When President Park assumed power, she brought back the MOF in 2013.

Furthermore, state agencies that are in charge of the oversight of vessel operations are multiple and poorly coordinated. For instance, the MOF oversees the operation of passenger ferries in coastal lines; the KCG is in charge of sightseeing boats in coastal lines; and local governments in cooperation with the National Emergency Management Agency oversee sightseeing boats in inland waters (Hankyoreh shinmun, April 24, 2014). Because these public offices operate under unclear, if not conflicting, administrative division of labor and under disintegrated regulatory regimes, the state response was delayed, clumsy, misinformed, and unorganized when the Sewol accident occurred. After the Sewol disaster, President Park asked the KCG to take responsibility for its failed rescue missions and ordered the dissolution of the KCG. Now there is a new ministry, the Ministry of Public Safety and Security, to centralize the management of public safety affairs.

Such constant reshuffling of cabinet ministries and state agencies makes it difficult to guarantee the responsibility and accountability of public authorities in their administrative performance. Under unclear division of labor and ineffective coordination, state agencies can simply avoid their administrative duties and blame each other for any failed functions. It also hinders the specialization and professionalization of public officers because such an administrative environment incentivizes individual survival within the organization, not professional specialization. Even the newly established Ministry of Public Safety and Security has constrantly been criticised for its inefficiency since its creation. This was closely related to President Park's decision to appoint former military generals with no professional expertise in public safety to lead the new Ministry and the Ministry's continued failures in subsequent disasters.

In addition to the problems of organizational disarray and under-professionalization, rescue missions have increasingly been privatized by being transferred from state institutions to private corporations. The Law regarding Disaster at Sea and Rescue was revised in August 2012 during the last few months of the Lee government to require the KCG to conduct rescue operations in cooperation with the Marine Rescue and Salvage Association (MRSA, *Haeyangkujohyeophoi* in Korean), a non-profit corporation (Hankyoreh shinmun, April 24, 2014). The MRSA was formed in January 2013 right after the revised law went into effect by private rescue and salvage companies associated with large shipbuilding and shipping corporations. As with any other private firm, their interest lies in making more profit, which comes from the frequency and scale of marine accidents. While the KCG is legally in charge of maritime rescue missions, the owner of the ship in accident is required by the revised law to contract a member company of the MRSA to do the real work of rescue and salvage. This stipulation does not only imply an extended degree of privatization in the operation of rescue functions but also provides an obvious opportunity for the collusion between the KCG and the MRSA.

In the actual situation of the Sewol accident, the KCG pressured Cheonghaejin to contract Undine Marine Industries for the exclusive monopoly of rescue missions of the Sewol (Donga Ilbo, October 7, 2014).[10] Undine Marine Industries is obviously a private company and a member of the MRSA. Despite the fact that another rescue vessel (Hyundai Boryeong) had already arrived at the accident location and that Undine's rescue vessel (Rivera) had not undergone safety inspections by the time of the Sewol's sinking, Cheonghaejin signed a contract with Undine to be in charge of the Sewol's rescue operation.[11] One of the fatal misconducts of the KCG was to offer Undine the exclusive access to the Sewol while prohibiting other rescuers—even those with proper equipment and professional training—from rescuing the passengers locked in the sinking Sewol during the most critical first several hours. In order to guarantee the exclusive operation of Undine, the KCG blocked the participation of its own special lifesaving units, the Korean Navy's Underwater Demolition Team, and private rescue professionals from the Underwater Environment Association, which all sought the KCG's permission to assist the rescue operation.

The collusive relationship between the KCG and the MRSA was obvious. Top officials of the KCG held top positions in the MRSA upon their retirement from the KCG. Among six paid executive officers of the MRSA, four were former KCG officials while more KCG retirees are involved in the MRSA in other capacities (OhmyNews, April 27, 2014). The chairman of Undine was the MRSA's vice president with

---

[10] Undine specializes in salvage operations and oil spill clean-up, not in passenger rescue and lifesaving.

[11] The bill that Undine submitted to the government for its 87-days of rescue operation was $1.6 million (Donga Ilbo, October 7, 2014).

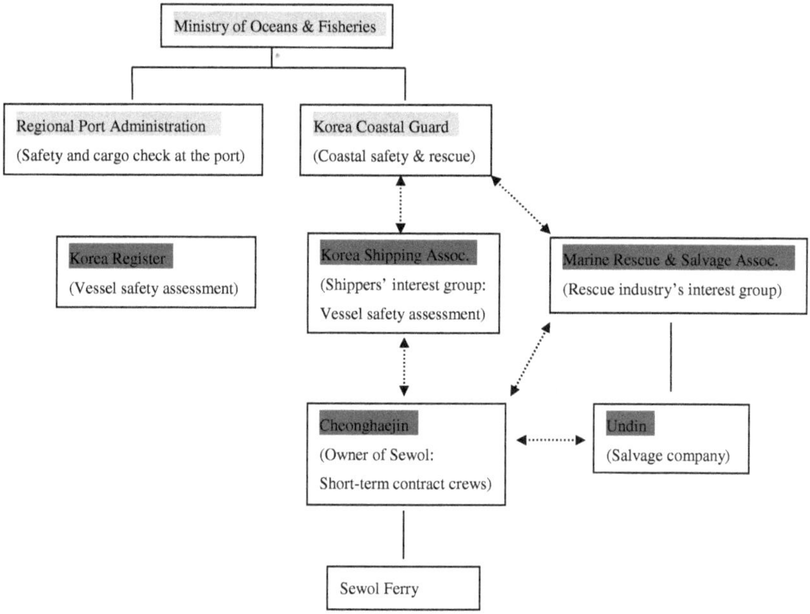

Fig. 2.1 Deregulation and collusion between the public and the private. *Notes* (1) Public institutions in *yellow*, (2) private entities in *light blue*, (3) organizational hierarchy in *straight lines* (___), and (4) collusive relationship in *dotted arrow lines* (……). *Source* Compiled by the author

close connections to the KCG. In the coastal transportation industry, it was widely known that KCG officials picked their favorite private companies when coastal accidents occurred (Kyunghyang shinmun, May 15, 2014).[12]

Figure 2.1 is a graphic summary of various public and private organizations involved in the Sewol disaster. This graphic summary shows the subdued function of public institutions in the management of public safety, the ascendance of private interest groups, and the collusive relationships between the public and the private, which altogether magnified the extent of the tragedy. Discussions on the political-economic conditions that lie behind the Sewol disaster as presented in this study indicate

[12] Undine made an income of $8 million (8 billion in Korean currency) for its rescue operation from April 16 to July 10, 2014.

that the Sewol's capsizing and the tragic death toll of 304 lives was not a simple accident caused by an unavoidable force of nature or by the misconduct of a few individuals. It was rather a predictable consequence born from the continued drive for deregulation and privatization in the area of public safety and the persistence of state–business collusion that has for long replaced strict oversight and transparency.

Concluding Thoughts

When debating the relationship between markets and states, even neoclassical economists acknowledge the role of the state in making interventions in areas of market failures because public goods are under-provided by purely profit-seeking economic actors. Public safety is obviously a public good that requires state provision and oversight. The Sewol disaster, however, demonstrates that the state increasingly withdrew its regulatory functions from the area of public safety and compromised its oversight responsibilities by allowing the ascendance of private interests in coastal transportation.

Yet, it is neither the purpose nor the intention of this study to present an opposition to every measure of deregulation or to claim that strict state regulation is a panacea to all sociopolitical ills. Deregulation itself is a loose concept applicable to a wide spectrum of policy reforms. Regulatory measures that hinder fair market competition or bureaucratic red tape that overburdens economic actors with unnecessary paperwork need to be deregulated. However, whenever there is state withdrawal, a form of deregulation, from the provision of public goods, we need to devise mechanisms through which private interests are incentivized to perform efficiently and professionally. This mechanism needs to include the provision of appropriate assessment and oversight by a third party. Whether the organization in charge of public safety is public or private, any form of unchecked interests accrues disproportional power to be used to promote corrupt collusions and blatant pursuit of self-interest.

The Sewol disaster and the complete incompetence of the government and state agencies in responding to disastrous emergencies pose critical questions about Korea's political structure and mechanisms of accountability. Korea is known for its strong state which is centered around an excessively powerful president. Traditionally, the incumbent president has made a strategic deployment of the executive branch to achieve his or her political agenda. Yet, as presidents are elected every five years and each new president operates within the limit of a single five-year term, administrative institutions undergo constant reshuffling. Under these repeatedly changing conditions, executive branches lack clear administrative division of labor, bureaucratic specialization, and political accountability. This implies that the capacity of the president and the central government to respond to emergencies and crises is severely undermined. Massive disasters like Sewol require swift and authoritative leadership with professional capabilities to make prompt intervention and to mobilize administrative coordination. However, the Sewol case demonstrated that a powerful president and a strong government were present only in political polemics, not in their actual performance for effective governing and management of the emergency response.

When the Sewol disaster is analyzed in association with a broad political-economic context as this chapter's exposition demonstrates, we begin to understand the magnitude this disaster poses to Korean society and democracy. The Sewol disaster reifies the fact that Korean society is loaded with a fundamental tension that lies in democratic capitalism or capitalist democracy. Democracy and capitalism operate on two distinctive and conflicting rules of resource allocation (Streeck 2011). Capitalism stands on economic rationality while democracy promises non-market notions of social equality.[13] Neoliberal capitalism in recent decades has exacerbated the conflict between market forces and democratic principles. In the process of embracing neoliberal shifts, businesses gained disproportional economic power and blatantly pursued their interests by translating their market dominance into political influence. Business interests in coastal transportation and marine salvage were able to expand their economic power and political influence to bring politicians and bureaucrats to align with their interests. They have lobbied politicians and bureaucrats to institute policies that would further advance their market expansion and profit accumulation. Politicians and bureaucrats receive kickback money and lucrative positions in the private sector in reward and become less and less accountable to the demands of

[13]Employers and the political right are profit-seeking actors that strive to undermine the majority rule of a democratic system because they fear that the property-less majority constantly demand the encroachment of private property and free markets through taxation and redistributive policies. The working people and the political left, on the other hand, strive to guard against capitalist interest that desires to undermine democracy by using its economic prowess.

the general public that brought them to power. Workers are treated as disposable goods that are squeezed to maximize employers' profit accumulation. Socioeconomic inequality widens and a growing number of the population is placed under material insecurity, which in turn weakens their political efficacy to influence political processes and policy outcomes. The genuine representation of the democratic majority becomes more and more infeasible and the mechanisms of accountability and transparency become more and more unreachable. This is the stark reality of how Korean society has been transformed in recent decades with the advancement of neoliberal capitalism.

Understood within this political and economic context, the Sewol disaster reveals the despicable abyss the Korean society has reached with the ascendance of capital interests. Neoliberalism has created a monstrous social norm that justifies the prioritization of corporate profits and monetary gains over any other values. In a neoliberal society where money rules, the guarantee of public safety through state intervention and public resources becomes a mere luxury. Under this extreme form of capitalism, democratic values such as equality, human rights, fairness, and transparency are relegated as naïve ideals, not actual practice.

References

Choi, Y., Kwangseok J., Cheolsoo L., & Beomsang Y. (2000). *Hankookeui nodongbeop kaejeongkwa nosagwangye* [Labor law reform and industrial relations in Korea]. Seoul: Korea Labor Institute.

Harvey, D. (2007). *A brief history of neoliberalism*. Oxford: Oxford University Press.

Kim, Y. (2012). *2012 bijeongkyujik kyumowa siltae* [The scope and reality of non-regular workers in 2012] Working Paper. Seoul: Korea Labor and Society Institute.

Lee, H.-w. (2012). Imyeongbak jeongbueui kyujegehyeok pyeongka [Assessing deregulation reforms under the Lee Myeong-Bak administration]. *Deregulation Studies, 21*(2), 3–38.

Lee, Y. (2014). Labor after neoliberalism: The birth of the insecure class in Korea. *Globalizations, 11*(4), 1–19.

Statistics Korea. (2013). *Kyeongje hwaldong inku chuga josa* [Additional survey of economically active population]. Seoul: Statistics Korea.

Statistics Korea. (various years). *Kyeongje hwaldong inku josa* [Survey of economically active population]. Seoul: Statistics Korea.

Streeck, W. (2011). The crisis of democratic capitalism. *New Left Review, 71*, 5–29.

Online Sources

Donga Ilbo: http://www.donga.com.
Hankyoreh shinmun: http://www.hani.co.kr.
KRS (Korean Register of Shipping): http://www.krs.co.kr.
KSA (Korea Shipping Association): http://www.haewoon.or.kr.
Kyunghyang shinmun: http://www.khan.co.kr.
Maritime Affairs and Port Administration: http://www.portincheon.go.kr.
PPSS: http://ppss.kr/archives/20978.
Pressian: http://www.pressian.com.
Redian: www.redian.org/archive/70395.

Author Biography

Yoonkyung Lee is Associate Professor in Sociology at the University of Toronto. Trained as a political scientist (Duke University 2006), she works on labor politics, social movements, and democratic institutions. She is the author of *Militants or Partisans: Labor Unions and Democratic Politics in Korea and Taiwan* (Stanford University Press, 2011) and journal articles that appeared in *Globalizations*, *Studies in Comparative International Development*, *Journal of Contemporary Asia*, *Asian Survey*, *Critical Asian Studies*, *Korea Observer*, and *Taiwanese Sociology*. Her next research project traces the historical formation of political opposition in Korea, equally shaped by a strong state and a vocal democracy movement.

CHAPTER 3

Institutionalized Irresponsibility: Understanding the Sewol Disaster from the Perspective of the Addictive Organization Theory

Su-Dol Kang

Introduction

Failed Theories

On April 16, 2014, approximately 304 lives were abandoned in the water while the Sewol ferry, going from Incheon to Jeju, was sinking. The official number of passengers was 476, of which approximately 304 lost their lives. Of the dead, 250 were high school students on a school trip. The lives were said to have been abandoned because: first, the captain and the crewmen, and then, the maritime police and the Navy that was responsible for rescuing the victims, and finally the government or the President that needed to take overall responsibility for ferry management or life safety, did not fulfill their proper duties.

S.-D. Kang (✉)
Korea University, Sejong, South Korea

J.-J. Suh and M. Kim (eds.), *Challenges of Modernization and Governance in South Korea*, DOI 10.1007/978-981-10-4023-8_3

Some view the Sewol ferry disaster just as a "traffic accident",[1] blaming the sinking on a combination of insufficient vessel safety checks and deteriorating weather. And they also say that the approximately 304 passengers that stayed in the cabins, following the onboard broadcasting and the crewmen's instructions, were not saved because of mistakes made by the crewmen and the maritime police, who were incompetent in the crisis. Also, the irresponsible attitudes shown by the captain and others who escaped first contributed to the results. However, these excuses hide the structural dysfunction long accumulated in Korean society, which reveal intellectual laziness, detachment, or lack of concern at a structural level.

Meanwhile, conspiracy theories surrounding the causes of the accident or the ex post facto response processes have been rampant in some areas such as social network services (SNS) like Facebook or Twitter. For instance, some argue that the scratches on the lower part of the stern of the Sewol ferry are traces of a collision with a submarine and that the accident was an intentional massacre. The guard ship of the maritime police, which was the first rescuer on the scene, did not rescue many passengers gathered near the stern. This guard ship rescued only a few people including the captain and an unidentified "orange man" from the stern. Some thought this guard ship hauled the Sewol ferry using ropes, and intentionally pulled it down.[2] According to this argument, there was some speculation based on information about illegal interventions by state organizations like the National Intelligence Service, the military, and the Ministry of Patriots-Veterans, and even ballot counting manipulation that the government needed a major incident to distract the electorate before the election. The Sewol disaster provided that distraction. However, there is a lack of objective and valid evidence to support these theories. Consequently, these views are generally regarded as rumors or conspiracy theories.

This article proposes a more comprehensive theory to understand the Sewol disaster. The Sewol disaster can be seen as a result of structural dysfunction in which the socioeconomic problems, accumulated in the process of economic growth over the last 50 years, came into focus.

---

[1] This "traffic accident theory" was originally presented by Joo, Ho-Young, a conservative ruling-party MP on July 24, 2014 and since then it has been pervasive among conservative civilians as well as politicians.

[2] For example, refer to Mediaus, September 3, 2014.

This structural dysfunction has been retained and reinforced rather than resolved because the chronic processes of addictive organizations have never been addressed or corrected. Addictive organizations are defined as those which act like active addicts. They deny, lie, control, manipulate, and forget (Schaef and Fassel 1988). At the time of writing, the truth of the disaster has still not been made public. Although 304 lives have been lost, nobody in positions of power is sincerely taking responsibility to try to make a system shift. This paper attempts to try to explore an appropriate logic or theory to clarify this unresolved problem. Furthermore, it hopes to develop a perspective that could be used to prevent repetitions of similar tragedies.

Theoretical Background

Social Responsibility of Corporations and the Government

Davis (1960) stated that Corporate Social Responsibility (CSR) is "decision-making of actions based on reasons that transcend corporations' economic or technical interest even slightly" and Eells and Walton (1961) defined CSR as "corporations' acts of not turning away their eyes from social problems and ethical principles that control the relationships between corporations and society." However, this definition of CSR was limited to economic and legal obligations, because only these obligations would secure the social legitimacy of corporate profit-making.

There are others who deny CSR. Friedman (1970) suggested that, given the freedom of market competition and guaranteed pursuit of profits, corporations would fulfill their proper responsibilities, including meeting legal and ethical obligations. However, this position assumes that consumers have perfect information and can make reasonable judgments. As long as that is not the case, legal and ethical problems are not solved through free-market mechanisms. Friedman also expresses the position that "Managers of corporations should pursue as much as possible profits while observing social rules embodied in given laws or customs in society," (Friedman 1970). Advising that corporations take economic responsibility, i.e., make profits for shareholders, while observing legal and ethical obligations, often is an inherent contradiction.

Other scholars argue that social responsibility should be incorporated in all the decision-making, policies, and acts of corporations. Carroll (1979, 1991) categorized CSR into a hierarchy of economic, legal, ethical, and charitable responsibilities such as supplying high-quality goods and services, creating jobs and social development, not circumventing the law, practicing civil society's ethics, norms, and expectations, and enhancing public welfare by supporting arts, education, or communities. Organizations that emphasize only economic responsibility tend to become corrupt, while organizations with robust ethical foundations show continuous financial growth. In Korea, many corporations are apt to neglect even the most basic legal and ethical responsibilities while using charity work for "impression management".[3]

With respect to governmental social responsibility, the idea in classical theories of social contract in which governments act to determine public interests to avoid chaos by enacting contracts in civil society were represented by Hobbes, Locke, and Rousseau (cf. Locke 1690). Some others argued, however, states are just agents of the ruling class to protect and enhance the rights and interests of the wealthy class (cf. Miliband 1969).

Today, most countries combine these two ideas as protecting people's lives and properties. Yet, protecting people's lives includes not only caring physically for them, but also must include enhancing quality of life and democracy, and being responsible for basic human rights. Protecting property also includes protecting socioeconomic systems based on private ownership, such as commercial corporate activities (Jessop 1990).

Perspective of the Addictive Organization Theory

Schaef and Fassel (1988) stated that organizations can demonstrate pathological behavior patterns. Schaef (1987) has also written that entire societies could become addictive systems. When an organization or a society shows unusual behaviors, we normally regard it as abnormal, but addictive organizations or addictive societies are often seen as normal or "reality". Addictions are often based on fear and the illusion of control. Alcoholics depend on alcohol to forget about pain and fear, get anxious

[3]For example, after the sinking of the Sewol, Samsung and Hyundai Heavy Industries, and Daewoo Shipbuilding & Marine Engineering provided marine cranes as humanitarian aids to support the rescue activities. Although the support had no actual effect for life-saving, their expense appeared as a charitable contribution. In addition, large corporations contributed to the Community Chest of Korea for the construction of national safety infrastructures, while separating this contribution from their everyday legal and ethical irresponsibility regarding unions or ecology.

without it, and never admit the existence of their addiction. They use falsehood and deception to hide their addiction and blame others for their "victimization."

The addictive organization is characterized as a dysfunctional system which acts in the same way as an active addict. According to Schaef and Fassel, there are four major ways through which an organization behaves like an addict.

First, if a core person of the organization—for instance, the CEO or other influential person in the organization—is an addict, the organization's decision-making processes and operational processes are likely to be dysfunctional. If the thinking or behavior of the significant person is not clear, the entire organization consumes time and energy trying to figure out what is happening. In addictive organizations, the role of codependents is crucial as they almost unconditionally support and comply with the thinking and behavior of the addict. Secondly, in an addictive organization, workers often carry their private behaviors into the workplace, which effects the entire organization. People who come from alcoholic families repeat the same dysfunctional patterns of their families in the organization. Thirdly, the organization itself acts as an addictive substance, as it offers unkept promises of possibilities and a sense of belonging. Also, workaholism becomes a powerful addictive substance to some people, supporting addictive behaviors. Lastly, organizations themselves can act as addicts. Communication is indirect and involves exaggeration, dishonesty, and concealment rather than honesty. Staff meetings avoid dealing with major issues thoroughly and discussion is limited and carefully controlled. Addictive organizations demonstrate symptoms of collective forgetfulness and do not and cannot learn from past experience or mistakes. They tend to blame others and do not own, examine, or correct their own mistakes. Irresponsibility is institutionalized and organizational thinking is characterized by either/or logic and rampant dualistic thinking. Addictive organizations are chaotic and when threatened by a crisis, frequently turn to opportunities to reinforce control through blaming others or finding a scapegoat. Addictive organizational structures reinforce formality, competition, control, and punishment, rather than ethical performance of the original mission. Both addicts and their codependents deny truth and responsibility and turn away from and delay resolving problems. My hypothesis is the Sewol ferry disaster was caused because all areas of Korean society, including politics, economy, culture, press, and education, resemble these addictive organization mechanisms.

Social Irresponsibility of Corporations and the Government in the Sewol Ferry Disaster

Root of the Sewol Ferry Disaster in the Light of Addictive Organization Theory

Confused Thinking and Frozen Feelings—Grandiose Goals, Dysfunctional Processes

If the government and civil society are healthy, regulations related to safety, health, environment, and human rights should not be relaxed, even under deregulation. However, addictive organizations lift all restrictions to focus on the survival and profit-maximization of the organization. For example, in the wake of neoliberal policy making, President Park Geun-hye said at a Cabinet meeting two months before the disaster, "I think regulations are enemies to defeat and cancer (?) killing us if not removed."[4] This way of thinking has bred societal pathologies like corruption, safety ignorance, and institutional irresponsibility during the last few decades.

Incidents similar to the Sewol disaster have often occurred in Korea and safety measures have been discussed every time. For instance, just two months before the Sewol disaster, ten people, including university students, died in the Mauna Ocean Resort tragedy in Gyeongju. Several days before this, the Minister of Security and Public Administration reported to the President that he would introduce a life-saving–golden-time system to make emergency vehicles arrive at the site quickly in an emergency. He emphasized that "the rate of arrival at the site within 5 min" would be enhanced from 58% (2013) to 74% (2017) to protect human lives. However, the reality was that the Park government just changed the name of the predecessor Lee Myeong-Bak's Ministry of Public Administration and Security into the Ministry of Security and Public Administration, without changing any content. On the contrary and to the public's loss, safety issues were assigned to the second vice

---

[4]Interestingly, right after the sinking of the Sewol, the Cabinet meeting passed an enforcement ordinance of the Housing Act that allows "vertical extension of apartments" from April 25, 2014.

minister's office, a non-influential department. The disaster management role assigned to the National Emergency Management Agency during the Roh Moo-Hyeon government was reorganized by the Lee Myeong-Bak and Park Geun-hye governments to fulfill presidential election pledges, with little effect.[5]

In addition, according to the MP Mun Byeong-Ho, although the Ministry of Maritime Affairs and Fisheries has an exclusive right to enforce the Marine Transportation Act, it has not exercised this right for the last 12 years (2003–2014). Article 60 of the Ministry of Maritime Affairs stipulates that "Violating acts on which fines may be imposed can be prosecuted only when the minister of the Ministry of Maritime Affairs and Fisheries has made an accusation." This statement indicates ignorance of safety procedures has become routine.

In fact, over the last 20 years, many great tragedies have repeatedly occurred.[6] Each tragedy was defined as a "man-made" calamity but all that followed was empty promises of paper measures such as safety-check systems or safety education. There have been neither punishments for those responsible nor thorough preventative measures instituted.

The Sewol ferry disaster also revealed the same problems. First, the vessel itself was 18 years old when bought from Japan in August 2012. It was supposed to have been scrapped two years later. However, in 2009, the Lee Myeong-Bak government extended the operable age of ships from 20 to 30 years because "The ships' age restricting system was too strict and unreasonably imposes burdens on shipping companies." For the same reason, the Park Geun-Hye government reduced field checks on business operators under the Ship Safety Act and allowed them to submit their own data instead of having an outside agency perform this

[5]Contrast this with the 1987 accident, in which the British passenger ship, the Herald of Free Enterprise, sank near Belgium and 188 people died. The authorities discussed how to "reinforce corporations' and business owners' responsibility for large accidents including industrial disasters". From this process, the Corporate Manslaughter and Corporate Homicide Act 2007 was enacted. Several million pounds of penalties were imposed on some corporations (Hankyoreh 21, May 9, 2014).

[6]These include: the sinking of the Yellow Sea ferry in 1993 (292 deaths), the collapse of Seongsu Grand Bridge in 1994 (32 deaths), the explosion of a subway in Daegu in 1995 (101 deaths), the collapse of Sampoong Department Store in 1995 (502 deaths), the fire in Sealand in 1999 (23 deaths), the fire in a subway in Daegu in 2003 (192 deaths), the Cheonan warship disaster in 2010 (46 deaths), and the Marine Corps camp accident in 2013 (5 deaths).

function. If the above was not enough to contribute to the irresponsibility of business owners, the law was amended to hold crews responsible for overloading or for excessive numbers of passengers.

In addition, safety checks on the number of passengers, cargo capacity, the number of vehicles on board, and ballast water was originally assigned to the maritime police pursuant to the Marine Transportation Act. However, instead of performing the work firsthand, the maritime police consigned it to the Korea Shipping Association (sailing control office), an organization representing shipping companies' interests. In the Sewol ferry disaster, it was revealed that the appropriate cargo volume was not observed and the number of passengers and vehicles on board (30 more cars were on board than recorded) was not checked accurately.[7]

Self-deception—Some Specifics About the Structure of the Sewol

Because of reconstruction, the center of gravity of the Sewol ferry was moved upward. The right side ramp weighing 50 tons was taken off and the fourth and fifth floors were added vertically to the stern. The prescribed number of passengers was increased from 804 to 921 through this addition. The result of this addition caused the hull to become asymmetrical and prone to severe shaking, and resulted in stability problems. This instability was so serious that it became a policy that "a tug boat shall be called even when winds blew."[8]

It is important to note that The Korean Register of Shipping was originally a non-profit corporation, which was entrusted by the Ministry of Maritime Affairs and Fisheries to manage ship safety issues. Nonetheless, from 2013, due to financial difficulties, it ignored its duty and began to pay attention to for-profit, non-ship-classification areas such as environmental technology tests and marine equipment evaluation. The Sewol ferry passed a ship registration test in February 2013

[7] The number of passengers was repeatedly corrected and even fatalities not in the list of passengers were found.

[8] In February 2014, the original Sewol captain Mr. Shin said, "The Sewol ferry is much more unstable than the Ohamana. The vessel shakes too much. Therefore, passengers complain greatly. This is because the ramp was taken off when it was remodeled after being imported from Japan." (KukminIlbo, April 23, 2014). A court testimony indicated that in early April, shortly before the accident, he said that "the vessel was the most dangerous in Korea" and that "it might cave over any time" (Segye Times, September 2, 2014).

without strict examination. It was approved for service until 2018, and was allowed to be put into service in the Incheon–Jeju route from March of that year.[9] In 2013, the Korean Register of Shipping company earned 120 billion won. Among its 12 past chair and head directors, eight were ex-government officials from the Ministry of Maritime Affairs and Fisheries, and many others were also ex-government officials. The Korea Shipping Association, which was also mandated to implement safety management, abandoned its mission and 240 out of 320 staff members concentrated instead on insurance businesses.[10]

In fact, documented problems such as improper or excessive loading always existed in the case of Chonghaejin Marine. For example, as much as three quarters of the ballast water of the Sewol ferry was taken out. In September 2013, Chonghaejin Marine was deeply concerned about safety problems and considered a resale of the Sewol ferry. The internal documents clearly reveal that Chonghaejin Marine was afraid of its asymmetric hull structure and vulnerability to strong winds. However, the company rationalized, "We tried to sell the vessel just to improve profitability, not because of its stability" (Sisain, May 17, 2014). Addictive behaviors that are based on a lie tend to produce other lies, as shown here. In short, although the Sewol ferry was an old vessel that should have stopped operating, its service life was crassly extended and the poor safety checks resulted in tragedy.[11]

The Addict Yu and His Codependents

When the Sewol ferry was being remodeled in 2012–2013, CEO Kim Han-Sik of Chonghaejin Marine and many staff members requested that

[9] When the vessel was being tested, Chonghaejin Marine submitted drawings different from actual remodeling and the Korean Register of Shipping tolerated the drawings while knowing that the drawings were false.

[10] "The governmental authority of the maritime police submits to powerful ship owners such as the owner of the Sewol ferry in many cases. I felt so sorry for the reality in which the maritime police could not exercise power very much even after seeing overloading or illegal cargoes." (Statement of a former conscripted policeman in the maritime police, October 11, 2014, personal interview.)

[11] In this respect, the conspiracy theory on social media indicating that Yu Byeong-Eon, an influential person of Chonghaejin Marine to which the Sewol ferry belonged, was intimate with former President Lee Myeong-Bak and funds for the 2012 presidential election came from Yu is put off for now. Of course, the banner that reads "Are we strangers?" that had been attached to the entrance of Geumsuwon in Anseong is highly symbolic.

they sell the ferry because of the issue of instability. However, chairman Yu Byeong-Eon, who himself behaved like an addict, disregarded the request.[12] Nobody raised the fatal issue again. Besides, he had even constructed his photo-exhibition room on the new fifth floor of the Sewol, ignoring some crewmen who said that this construction raised safety and instability problems.[13] Why did no one pursue these instability concerns? It appears that fear of exclusion prevented crewmen or staff from further raising safety issues with the company.

In addition, it seems that Yu's family not only participated in illegal tax evasion, breaches of trust, and embezzlement, he also strongly lobbied National Assembly members for an amendment of the Marine Transportation Act which would be favorable to his company.[14] Consequently, in addition to extending the vessel's service life, the Lee Myeong-Bak administration relaxed the engine test requirement from every 7000 h of engine operation to every 9000 h. At the same time, the government allowed the shipping company to buy an 18-year-old vessel, dangerously rebuild it, and make money with it for at least five years. Every year, Chonghaejin Marine gave a least one billion won of slush funds to Yu's organization.

Yu himself led the Evangelic Baptist, or Saviorist, Church. He combined religion and business in the 1970s, establishing the Semo Group in 1982 and launching a Han River cruise ship business in 1986. However, the Semo Group went into receivership in 1997. Then, he used the receivership to take over the group through a shell company and established Chonghaejin Marine in 1999. After failing to pay 200 billion won at that time, he later accumulated properties amounting to 560 billion won in 10 years through "phantom management." Most executives and staff members of the 15 affiliates remotely controlled by him were Saviorists. It seems that staff members' employment was determined more on testimonies rather than job abilities.

[12]Chairman Yu ignored this proposal and instructed the sale of the Ohamana known as "a twin vessel," thinking nothing of the problems in restoring force of the Sewol and further encouraged habitual overloading.

[13]Three elements important for maintaining vessels' balance are: vessel structures, ballast water, and cargo loading.

[14]Money for lobbying is said to have been procured by selling photos taken by Yu under Ahae. Eventually, it is estimated that at least 20 billion won of slush funds were raised through more than ten photo exhibitions.

As shown in Table 3.1, Chonghaejin Marine is ranked in the middle for sales, while the a mounts spent on education, pay, and welfare, which correspond to legal and ethical responsibilities and charitable contributions, are extremely low.[15] Changes in expenditure structures in the company are more surprising. The contributions of Chonghaejin Marine to charities increased from 300,000 won in 2010 to one million won in 2011 but decreased to 500,000 won in 2012 and 300,000 won in 2013. In addition, educational and training expenses, including for safety training, decreased to 540,000 won in 2013.

Relatively poor treatment of employees results in employees who lack organizational commitment and job involvement. The Captain, Lee Jun-Seok, had retired and then become the Sewol's substitute captain as a contract worker at the time of the disaster. The original captain, on vacation at that time, said that he had repeatedly warned that the vessel was dangerous whenever he sailed to Jeju, but his opinion had not been listened to or accepted by his CEO. Another contributing factor was that the monthly wages of the mate, chief engineer, and engineer were around 60–70% of other comparable shipping companies' salaries. In addition, 9 out of 15 ship personnel were contract workers. These conditions further contributed to the dysfunction of the organization of the vessel.

Collusive Relations and Moral Deterioration

In addition, Chonghaejin's reception expenses, which are often linked to corruption, are mid-ranked in the industry. These expenses increased greatly from 50.7 million won in 2012 to 60.57 million won in 2013. Reception expenses were approximately 200 times the amount of charitable donations in 2013 (cf. Table 1). In short, whereas legal, ethical, and charitable responsibilities decreased from 2012 to 2013, social irresponsibility increased disproportionally.

Interestingly, Chonghaejin Marine engaged in CSR when it took patients with Hansen's Disease on a trip to Jeju Island in November 2005. However, at the same time the company, as usual, was overloading passenger ships, not providing sufficient ballast water, ignoring

[15] Of course, the contributions might be bribes rather than pure contributions. According to the table, entertainment expenses fall into the latter category.

Table 3.1 Comparison of expenditures—nine largest shipping companies in South Korea (2013)

Expenditure (1000 won)	*Goryeo*	*Namhae*	*Daea*	*Dongyang*	*Bugwan*	*Sea World*	*Chonghaejin*	*Panstar*	*Hanil*
Sales (bn. won)	6.69	17.2	45.6	0.33	19.9	53.3	32.0	46.5	49.2
Education	569	0	11,139	280	0	1596	541	13,771	0
Average pay	43,123	30,856	70,532	39,376	126,881	43,585	36,331	49,337	38,334
Av. welfare	5715	3139	15,242	6360	17,395	5145	1410	12,455	5249
Donation	18,438	10,000	43,780	12,500	0	153,890	300	53,592	10,000
Entertainment	33,034	44,907	201,688	17,777	208,932	115,943	60,574	401,265	82,070

Data Referring to shipping companies' audit reports, Economic Review (April 24, 2014). The author reconstructed the data

The average pay and welfare expenses are amounts per person and other items are total amounts per company

Goryeo = Goryeo Express, Namhae = Namhae Express, Daea = Daea Express Shipping, Dongyang = Dongyang Express, Bugwan = Bugwan Ferry, Sea World = Sea World Express, Chonghaejin = Chonghaejin Marine, Panstar = Panstarline.com, Hanil = Hanil Express safety checks and safety warnings, and exploiting non-regular workers.[16] The company's employee safety education expenses in 2013 were only 540,000 won, and some employees had never been educated in safety practices after they started their employment. The Sewol's captain originally submitted a report to the sailing control office indicating that 150 vehicles and 657 tons of cargo were loaded. Chonghaejin Marine announced in the very afternoon of the disaster that 180 cars and 1157 tons of cargo were loaded. However, it was finally revealed that a total of 3600 tons of cargo were loaded, although the carrying capacity was only 987 tons. As many as three 50-ton trailers were on board, but not properly secured.

Despite its background as a specialized salvage company, not a rescue company, Undine, a search-and-rescue company under exclusive contract to Chonghaejin Marine, was guaranteed exclusive access to search-and-rescue activities. Undine is currently an unlisted company but was planning to be listed on the stock market in 2014–2015. In fact, 36% of its shares are owned by an investment association comprising the Ministry of Strategy and Finance, the Ministry of Land, Transport and Maritime Affairs, and the Korean Intellectual Property Office.[17] Undine's personal connections with high-ranking officials are many. It has already received at least 300 million won of governmental subsidies and approximately 4 billion won of public payment guarantees. In 2011, it implemented stone paving work for fixing undersea cables between Jindo and Jeju provinces and made approximately 1.2 billion won profits out of the total construction price of 5 billion won. This project was originally ordered by Korea Electric Power Corporation and given to Deokman Shipping Co. through LS Cable & System, and then subcontracted to Undine. However, Deokman Shipping Co. is a shell company and there is evidence of its receiving slush funds.[18] If the Sewol ferry disaster had not occurred, Undine could have monopolized the project to salvage approximately 1800 sunken vessels in the littoral sea of the Korean peninsula.

---

[16]Currently, out of 802 crewmen of coastal passenger ships in Korea, 602 (75%) are contract or non-regular workers (Ohmynews, July 31, 2014).

[17]The 10th session of Kim Eoh-Jun's Papais, 'The *Sewol* ferry, *Undine*'s greed', June 20, 2014.

[18]Op. cit.

According to its audit reports, the amount of charitable contributions of Undine in 2013 was 55.70 million won. There was no contribution in 2012. This suggests that there is a possibility that Undine bribed some officials when it joined the Maritime Rescue & Salvage Association (MRSA). The MRSA was formed by ruling and opposition party National Assembly members in January 2013 under the auspices of the Sea and River Disaster Relief Act, amended in 2011. The positions of president and vice president were taken by the chairman of Sewoon Steel and the president of Undine, respectively. Six retirees from the maritime police were appointed to the MRSA. Vice-chairman of the National Assembly, Lee Byeong-Seok, who representatively proposed the bill at that time, was an adviser to the MRSA. Former chief Lee Gang-Deok of the maritime police is an honorary president. The National Assembly, the maritime police, and Undine were brought together in a single body.

The collusion extends further. For instance, the government-run Korea Development Bank lent 16.9 billion won to Chonghaejin Marine. In October 2012, it lent the company 10 billion won in security for the purchase of Sewol ferry despite its high debt ratio.[19] If the company's soundness or risk value had been properly analyzed, that amount of loan should have never been possible. Using 3 billion won out of the loan money, the company renovated the vessel to increase its capacity by 239 tons and the prescribed number of passengers by 116. According to the Financial Supervisory Service, the balance of eight Semo Group companies' bank loans at the end of 2013 was 137 billion won and the debt from the entire financial world was 200 billion won. In particular, two companies, Chonghaejin Marine and Chonhaeji, received 30% of their loans from the Korea Development Bank and the Industrial Bank of Korea.[20] This can only be explained by assuming the mediation of high-ranking government officials, which strongly suggests government-controlled financing and collusive links between politicians and businessmen.

Out of the Sewol's 46 lifeboats, 45 were not unfolded during the accident. Since March 2013, the Korea Shipping Association and the maritime police supposedly conducted safety checks six times on the Sewol, in which cargo fixing devices and lifeboats were indicated to be

[19] The Seoul Finance, April, 21, 2014.

[20] Other debts included 870 million won to Kookmin Bank, 800 million won to Shinhan Bank, 1 billion won to Hana Bank, and 1 billion won to Korea Exchange Bank (Seoul Finance, April 21, 2014).

"no problem" or "good." It seems that it was not adequately checked whether they operated or not. Emergency training of crewmen was judged "insufficient" in June 2013 and "no record" in August 2013 by the maritime police but was judged "good" in November, despite no action having been taken. According to Newstapa, one of the influential independent media outlets, officers took only 13 min to check the safety of a 6825-ton ship. The KRS entertained officials of the Ministry of Maritime Affairs and Fisheries using corporate credit cards for drinking and golf. Both workaholism and alcoholism were burgeoning through corrupted relationships (Fig. 3.1).

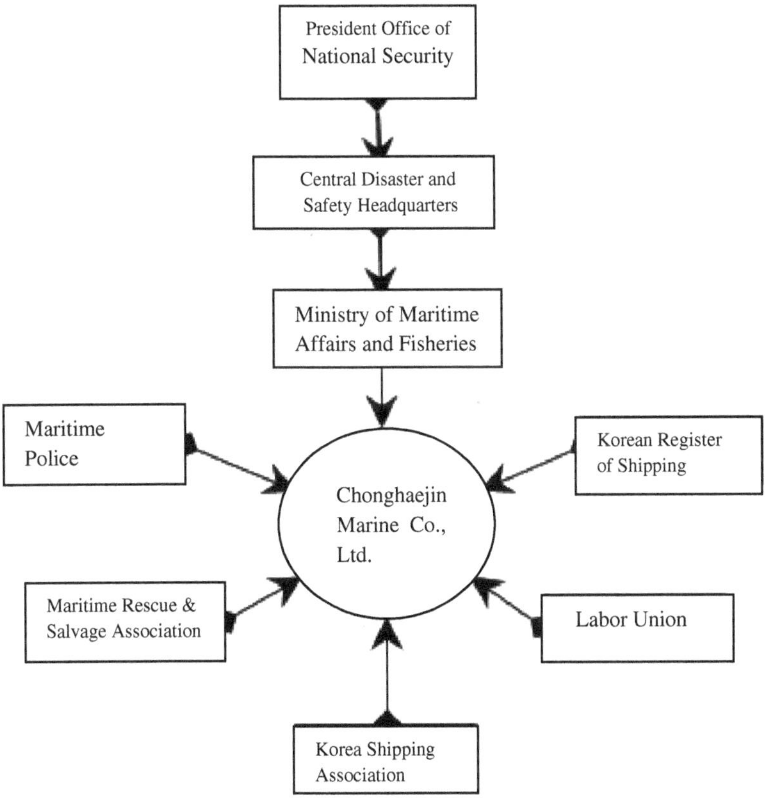

Fig. 3.1 Relationships among organizations surrounding Chonghaejin Marine Co., Ltd. *Data* Reconstructed by this author referring to Sisain, May 24, 2014

In short, members of the National Assembly and the "sea mafia" have been forming mutually addictive relationships for money-making rather than maritime safety or ecological preservation.

The role of the labor union is another aspect worth noting. Healthy unions should have a sense of responsibility, not only for workers' rights but also for passenger safety. In the case of the Sewol ferry, there was a port worker who mounted a one-person demonstration to protest against habitual overloading.[21] However, both the company and the port transportation labor union ignored his protest. It appears that acts of pursuing justice or responsibility are substantially ignored or actually forbidden in corporations and society. Those who pursue justice are easily branded as interfering with business and excluded from participation. If there had been a proper union in Chonghaejin Marine to which the Sewol ferry belonged and non-regular crewmen actively participated in that union, and if workers' representatives could have participated in the board of directors, this disaster could have been prevented.

Addictive Processes in Ex Post Facto Responses for the Sewol Disaster

Habitual Expediency and Organizational Incompetence

At the moment the Sewol ferry was sinking, despite being obliged to remain on the vessel until the end, the captain and the crew escaped first. According to sailing control regulations, their role was to manage the situation and rescue passengers. Instead, they instructed the passengers to "stay still" several times and neglected their proper rescue duties.

It is reported that the staff members of Chonghaejin manipulated the Sewol ferry's cargo book because they judged "that the cause of the accident was the damage to restoring force" and tried urgently to "hide the fact of overloading".[22] In addition, they were concerned that insurance money would not be received if overloading was identified as the cause.

[21] Jeong (2014).

[22] According to the Financial Supervisory Service, the Sewol ferry is insured by Hull Insurance for 11.3 billion won. Meritz Fire & Marine Insurance Co. undertook 7.8 billion won and Korea Shipping Association undertook 3.6 billion won (Korea Insurance Newspaper, April 27, 2014).

In fact, the Sewol ferry sailed the Incheon–Jeju section 241 times since March 2013 and the vessel was overloaded 239 times. The load at the time of the accident was about 3600 tons, at least three times the appropriate load. Immediately after the accident, staff manipulated documents as they were instructed to do by their superiors. Unfortunately for management, they only reduced the cargo volume by 180 tons, because none of them knew the appropriate capacity. This focus was a way for the addictive organization to survive at a moment when more than 300 lives were threatened.

It is still not clear whether the reported "rapid veering" was a direct cause of the disaster. The only information presented thus far is the court testimony on June 10, 2014. The third mate, Mr. Park, of the Sewol ferry stated, "One vessel came up on the opposite side" and "I instructed the tillerman to veer within five degrees as usual in order to avoid a collision" (Seoul Daily, June 12, 2014). Authorities should have disclosed the identity of the vessel. However, this has not yet happened.

It appears that investigative authorities like the prosecution were not interested in discovering the truth. Instead, in response to online rumors, they just warned that "False rumor canards will be punished severely." (Ohmynews, May 1, 2014). The authorities also interfered with accurate disclosure. For example, the contents of communication messages were cut out and radar images and Automatic Identification System records damaged.

Even against the request for establishment of a special law for truth ascertainment, some obscure the essence of this request by referring to "special admission into university" or "compensation issues" or even block the discourse by saying, "Now we are fed up, stop speaking about the Sewol ferry." (Hankyoreh, September 10, 2014). In fact, the idea of "special admission into university" or "compensation issues" came from the ruling party, so that the burden of the government with respect to the disaster can be relieved. Victim's families, however, really wanted the truth. While the authorities function like addicts, a large part of society plays the role of their codependents.

Codependent Relationships Resulting in "Golden Time" Being Lost

Meanwhile, according to Chonghaejin Marine, the maritime police sent a fax message at about 2:00 pm on the day of the disaster to Chonghaejin to instruct them to erect a marine crane at the site[23] and threatened to charge the company with the expense. Later, the maritime police urged Chonghaejin to sign a contract with Undine because Undine was already performing rescue work. Interestingly, on the next day, Undine visited Chonghaejin with a contract that read, "The parties agree that Undine will exclusively perform all rescue and salvage work and other technical support etc. for the Sewol ferry and make a service contract as follows."[24] The contract money or amount of compensation was not specified. The prices were "all up to Undine." The marine crane departed for the disaster site 12 h after the accident and arrived at the accident site 55 h later, but the vessel had completely sunk before the arrival of the crane.

Ideally, the maritime police should play a central role in rescue and salvage activities, the Blue House should control operations, and the Ministry of Public Administration and Security or the Ministry of Maritime Affairs and Fisheries should be intermediaries, providing necessary support. However, as is documented, the maritime police intermediated an exclusive contract between Chonghaejin Marine and Undine, refusing to "firsthand call a marine crane because of money."

In addition, in an unusual move, an investigator of the maritime police helped the captain escape from the vessel and took him to his private home isolating the captain from reporters and other investigators. This behavior is contrasted with the case of the sinking of the West Sea ferry in 1993 in which the maritime police were in charge of the overall

[23]Chonghaejin Marine did not know about Undine even when the disaster occurred. A staff member of Chonghaejin said, "The maritime police officer in charge said, 'Although I am a little embarrassed myself to say this, there is a company named Undine and this company has been already performing salvage work. Make a contract with that company.'" (Sisain, May 10, 2014).

[24]The maritime police and Undine performed missing person search operations at a cost of 500 million won after the sinking of the vessel Geumyang that was searching for missing soldiers in the Cheonan warship disaster in 2010. Undine received 450 million won from the government at that time, despite the fact that there was no outcome at all (Kyunghyang Shinmun, April 27, 2014).

investigation of the accident and were dedicated to lifesaving and the proper prosecution.[25] As it turned out, the investigator in the Sewol disaster was found to be a member of Yu Byeong-Eon's Saviorist sect.

The Mokpo maritime police that received the captain's calls and rushed to the site did not try to prevent the vessel from sinking or to rescue passengers. The police rescued only ten crewmen who had made improper announcements and then conducted rescue activities without entering the vessel, saving those passengers that plunged into the sea. The maritime police even lied to the press that "the passengers were being rescued by 555 rescuers, 121 helicopters, and 169 vessels." However, that evening, the on-site family members of the missing persons disclosed that the number of rescuers was smaller than 200 and only two helicopters, two warships, two guard ships, six special-forces boats, and eight private rescuers were seen. It would appear that the maritime police did not rescue any victims. The fact that the maritime police neglected their duties and then took 50 out of the 68 assigned positions in the initial joint investigation headquarters is a clear act of deceit. About 2 h later, even the press released a false report indicating "all passengers were rescued." It seems that the press was not interested in reporting the truth or investigating the causes but only competed for sensational coverage. All of these behaviors are indicative of systemic addictive behaviors and functioning.

Codependents to the Key Person

After the disaster happened, the owner of Chonghaejin Marine, Yu Byeong-Eon, remained at large. Some subordinates like Mother Kim, Mother Shin, and driver Yang, the Saviorist believers, helped Yu with his escape and concealment, until his corpse was found on June 12.

However, Yu is not the main cause of the "rapid veering." Nonetheless, the authorities and the press acted as if he were. Shifting blame is another characteristic of addictive behavior. The fact that the authorities kept silent after discovering Yu's death, that they declared his

[25] In this incident that occurred in 1993, the maritime police organically cooperated with private fishing boats in lifesaving and dead body retrieval from the beginning to rescue 70 out of 362 passengers (292 died). Interestingly, the white paper released later, however, pointed out the same problems as in the Sewol case, such as an unreasonable departure in bad weather, unreasonable veering, an excessive number of passengers, overloading of cargo on the deck, and poor-quality lifeboats (Sisain, May 17, 2014).

death accidental after one month and conducted his funeral quietly on August 31, was abnormal in Korean culture.

In contrast, Chonghaejin Marine refused to pay funeral expenses for two part-time workers who died and two other part-time workers who were lost. Its position was that funeral expenses cannot be paid because part-time workers did not join the mutual aid fund.

The executives and staff members of Chonghaejin Marine did not sincerely apologize to bereaved families in the process of trials. They tried to blame the captain and the crew. The Blue House, the President, and the government said they would take full responsibility by establishing a special law to determine the truth, punish those responsible, and prevent recurrence but showed later that they only had an attitude to wrap up the situation with empty promises.

Pretense in Truth-Finding

The court in Gwangju decided in November 2014 that Captain Lee was guilty of negligence, but not of homicide, for which prosecutors had sought the death penalty. Lee was jailed for 36 years. The court convicted the ship's chief engineer of homicide for not aiding two injured fellow crew members. This conviction did not address the 304 lost passengers. They sentenced him to 30 years' imprisonment. The remaining 13 surviving crew members of the Sewol ferry were found guilty of various charges, including negligence, and were imprisoned for 5–20 years.

These verdicts and sentences, however, were inadequate in finding out the substantial truth of the disaster and so fulfilling the wish of victim's families to dispel all the suspicions raised after the tragedy. Moreover, such a judicial decision can narrow our sight and misdirect our attention, as the legal responsibility of the crew is just the tip of the iceberg and they are probably not the real criminals. As a result, these kinds of decisions can shift people's attention from the essence of the responsibility that dysfunctional corporations and the government or the whole society should face.

And even the Special Sewol Law[26] passed at the National Assembly on November 7 after a long struggle did not guarantee that the whole truth of the disaster was uncovered and that those substantially responsible

[26]The Hankyoreh, November 8, 2014.

were identified and punished. The loophole in the law lay in the fact that the special committee was entitled neither to investigate those responsible nor to charge them. Even at the hearings in December 2015 organized by the committee after another prolonged struggle, those responsible did not show up or answered dishonestly. For instance, most witnesses responsible for the disaster either said, "I don't know" or "I don't remember exactly." And the CEO of Undine and the president of the MRSA did not attend the hearing for "unreasonable grounds." The special committee decided finally to accuse these two of being unfaithful as witnesses (SBS, January 11, 2016). In this process, both the government and the ruling party have tried to avoid the responsibility to reveal the truth and to punish those responsible.

Repeated Dishonesty

Interestingly, a document titled "comments of the National Intelligence Service" found in the Sewol ferry notebook computer was restored on July 25, 2014. This document was originally created at the end of February 2013 when the extension and the alteration of the Sewol ferry were being finished. It shows instructions on about 100 min details including refrigerators, TVs, and painting. This document created a reasonable doubt whether the actual owner of the Sewol ferry was the National Intelligence Service instead of Yu. In addition, the "marine accident reporting system" in the Sewol ferry's sailing management regulations required the Sewol ferry to first report accidents to the Jeju and Incheon branches of the National Intelligence Service. However, the National Intelligence Service provided explanations that did not account for anything. These illogical explanations strengthened the reasonable suspicion that the government was complicit in the lies or was concealing further information.

Meanwhile, in the vortex of the social sorrow and anger about the Sewol ferry tragedy, some corporations tried to utilize what should have been time for mourning as an opportunity for moneymaking. For instance, K Sport shop advertised its discount event while posting letters of mourning for the victims on an online community. And the G-Bookstore prepared a disaster-related, book-selling promotional event. These corporations obviously displayed addictive behaviors by treating a tragedy as a moneymaking opportunity as they had lost touch with the meaning and importance of the process of mourning in themselves and others.

The government and the Blue House were the worst. Some governors visited Paengmok port in Jindo and took commemorative photographs after offering formal consolations, eating Ramen noodles near the missing persons' families. On April 23, Kim Jang-Su, the head of the national security office of the Blue House, even declared that the Blue House was not the control tower of disasters.[27]

In addition, President Park visited the Jin-do Gymnasium one day after the disaster and said, "Since an incident that cannot happen happened, I will thoroughly investigate and ascertain the cause and will severely punish all those deemed responsible." As the person ultimately responsible for the system, she took the viewpoint of a third party. In the same way, on April 21, she unilaterally reprimanded ministers and vice ministers and defined the captain and crewmen as "murderers," without apologizing for her part in the disaster as a recovering addict would have done.

Conclusion: "System Shift" for Recovery of Social Responsibility

Like individuals, addictive organizations and addictive societies can also act without any sense of responsibility for their behavior. The results of addictive behavior appear as social irresponsibility in an institutionalized form. Corporations privatize profits and socialize costs, while the government attributes criticism to others, makes big promises and practices dysfunction.

Le Monde in France (April 19, 2014) commented, "What is the use of the fastest Internet in the world, the best smart phones, and excellent shipyards if they cannot rescue their own children?" For over more than 50 years of industrialization, Korean society has ignored basic commitments to health and quality of life, only to get addicted to quantitative growth, external appearances, and being the best. Being acclaimed as the

---

[27] However, in fact, the manuals of the maritime police and the Ministry of Maritime Affairs and Fisheries set forth the national security office of the Blue House as the control tower (Nocut News, April 24, 2014). Besides, MP Kim Hyeon-Mi found out later that the vice ministers of diplomacy, national defense, and national unification and the National Intelligence Service assistant chief gathered in the Blue House at the very time of sinking of the Sewol ferry and held a "National Security Council (NSC) meeting" for 1 h from 8:30 am, and that they allegedly did not know about the Sewol ferry accident (Pressian, July 10, 2014).

"Asian dragon" is the active drug in this addictive system. In short, the Sewol ferry tragedy paradoxically offers an opportunity for reflection on the development of this kind of society.

Like individuals, an addictive society should admit that it has an addictive process from which it needs to recover. However, since in an addictive society almost all members are actually codependents, health can be recovered only when all members of society sincerely participate in the process of healing.

Healing means that the problem cannot be solved by conducting more thorough audits, preparing CSR reports, or holding a ceremony to declare "ethical management." It is not enough for corporations to create employment by investing capital, satisfying customers' needs, and achieving economic growth, just to be measured on the yardstick of legal and ethical responsibilities. That is to say that all the pathological elements that dwell in all members, organizations, administrative processes, and economic and political processes should be honestly faced and named as complicit with the disease of addiction.

We are addicted to materials such as money, alcohol, and commodities as well as to processes such as work, power, relationships, consumption, SNS, games, and even self-abuse, self-centeredness and the illusion of control. The difficulty in finding the truth of the Sewol ferry disaster is deeply embedded in the fear of the loss of money, power, status, or belief. The real problem lies in the very obsession that one could solve all problems with money, power, and economic growth.

The core of the addictive system surrounding the Sewol ferry was embedded in a structure in which a vessel that was about to be scrapped was legally approved even after being unreasonably extended. It was habitually overloaded, poor safety checks were tolerated, and the system in which it functioned facilitated alliances that existed not only between Chonghaejin and Undine but also among the maritime police, the Korea Shipping Association, MRSA, the National Assembly, the government, the National Intelligence Service, and the Blue House.

To create a healthy society, a sense of responsibility for life should be recovered. Once healing has begun, the person or persons responsible should have a healthy attitude of responsibility to declare the beginning of healing and consistently lead organizational and societal innovation and to confess his/her lack of personal responsibility and capacity. Only then can the process of healing begin in the entire society. As such, corporations and the government need to be reborn with a sense of social responsibility and participate in the process of system shifts from sick to healthy organizations which can occur in the entire society (Schaef 1987; Schaef and Fassel 1988).

To achieve this process of system shifting toward healthy organizations, first, we should honestly admit we are addicted to money, power, work, status, relationships, and many other things.

Second, individuals and organizations should stop participating in addictive processes inherent in their very structure. For example, freedom of association in this sense should be guaranteed and extended in all areas of society to correct the "absolute imbalance of social power."[28] To put it another way, healthy organizational structures and operating methods, in which diverse voices are freely heard and reflected upon with respect to policies and practice through processes of earnest discussions, can be a breakthrough.

Third, all the processes of organizations (mission performing, communication method, decision making, compensation structure, value sharing, skill training, and working style) should be innovated consistently.

Only when CSR and the government's social responsibility have been recovered through this system shift can the entire society become healthy and enhance the quality of everyone's right to a meaningful and fulfilling life.

References

Carroll, A. B. (1979). A three-dimensional conceptual model of corporate social performance. *Academy of Management Review, 4*(4), 497–505.

Carroll, A. B. (1991). The pyramid of corporate social responsibility: Toward the moral management of organizational stakeholders. *Business Horizons, 34*(4), 39–48.

Choi, J. (2014, June 24). Sewolho sagunul tonghebon chegimeui sahoijuk jogun [Social conditions for responsibility seen through the *Sewol* ferry incident]. *The Open Platform* (online). Seoul: Naver.

Davis, K. (1960). Can business afford to ignore its social responsibilities? *California Management Review, 2*(3), 70–76.

Eells, R., & Walton, C. (1961). *Conceptual foundations of business*. Homewood, IL: Richard, D. Irwin.

[28] Choi (2014) and Weekly Kyunghyang, September 16, 2014.

Friedman, M. (1970, September 13). The social responsibility of business is to increase its profits. *New York Times*, 122–126.

Jeong, S. (2014, May 23). Boiji aneun soni je ieui chamsarul ilukinda [The invisible hand will cause the second *Sewol* ferry tragedy]. *Pressian.*

Jessop, B. (1990). *State theory: Putting the capitalist state in its place*. Cambridge: Polity Press.

Le Monde Corée du Sud : arrestation du capitaine du ferry naufragé, April 19, 2014.

Locke, J. (1690). *Second treatise of government* (10th ed.). Project Gutenberg. Retrieved March 25, 2012.

Miliband, R. (1969). *The state in capitalist society*. London: Weidenfeld and Nicolson.

Schaef, A. W. (1987). *When society becomes addict*. San Francisco: Harper.

Schaef, A. W., & Fassel, D. (1988). *The addictive organization*. San Francisco: Harper.

Author Biography

Su-Dol Kang is Professor at Korea University in Chochiwon, Sejong-City. He received Dr. rer. pol. from the University of Bremen, Germany, majoring in industrial relations. His research focuses on labor relations, work addiction, addictive organization theory addictive society, and critical analysis of management practices. His major publications include *Fordismus und Hyundäismus: Rationalisierung und Wandel der Automobilindustrie in der BRD und in Südkorea* (Peter Lang Publishing, 1995), "Typology and Conditions of Migrant Workers in South Korea," *Asian and Pacific Migration Journal*, 5 (1996): 265–279, and "Labour Relations in Korea between Crisis Management and Living Solidarity" *Inter-Asia Cultural Studies*, 1 (2000): 393–407, among many others.

CHAPTER 4

National Crisis and Democratic Consolidation in South Korea

Taehyun Nam

A country's institutional capacity, leadership, and legitimacy are put to test during times of national crises, and the government's missteps could shake up the facade of normalcy and legitimacy. Such an incident occurred in South Korea in 2014 when the public's suspicions toward the government deepened to the point where many began to question the very legitimacy of the government. On April 16, the Sewol ferry capsized about one mile off Donggeochado, South Jeolla Province. Soon the ferry began to sink helplessly, endangering the hundreds of lives on board. The South Korean public turned to the government for help, and was given the impression that an immediate and successful rescue mission was under way. News reports even declared at one point that everyone was rescued. However, within hours, the public learned that out of 476 passengers—the number that the government could not verify for many days—fewer than 200 were rescued, mostly by fishing boats. Even President Park Geun-hye mysteriously disappeared from the public for seven hours on the day of the disaster. As the days went by with no additional rescue, the public became increasingly enraged with the

T. Nam (✉)
Salisbury University, Salisbury, USA

J.-J. Suh and M. Kim (eds.), *Challenges of Modernization and Governance in South Korea*, DOI 10.1007/978-981-10-4023-8_4

government's rescue mission that was disoriented, ineffective, and largely useless. As turned out later the South Korean government's safety regime had turned a blind eye to the illegal restructuring of the boat as well as cargo overload. The Cost Guard missed the crucial opportunity to rescue passengers and even refused assistance offered by the military and other agencies. The government manipulated the mass media to hide their mistakes (PSPD 2014). The public anger was compounded since most of the 300 or so dead were high-school students from the same school. To this day, the government's investigation of the incident remains incomplete and prosecution of the responsible parties was largely limited to the sailors of the ferry, leaving many corrupt and irresponsible government agents unaccountable. The families of the victims protested, many starting hunger strikes against the government, and public support has been enduring and passionate.

As the public's anger turned to disillusionment with the regime, many began to wonder about the degree of the country's democratic consolidation that many believed completed. The watershed moment of South Korean democracy came in 1987 when President Chun Doo-hwan gave into the demands of the opposition and allowed a direct presidential election. Throughout the country, citizens flooded the streets, cheering for their victory and with grand hopes for a democratic future. Such optimism is not rare among newly democratized states, as witnessed after South Africa's free and fair election in 1994 or Afghanistan's presidential election in 2005. These hopes are not just about the governance but stem from notion that democratization will have positive effect on other areas beyond politics. Even scholarly discussions of democracy tend to covertly assume, if not sometimes overtly state, that democratization is benign and benevolent. For example, Lijphart (1993) argues that democratization would change the social norms of "representativeness, accountability, equality and participation" (p. 149), which in turn would further strengthen democracy. Some see growing accountability as a positive consequence of democratization (Locke 1924) or even a power shift towards the people (Dahl 1989) through increased public participation in governance (Mill 1991, pp. 277–279). Democratization is also a promise of civil rights and liberty on an unprecedented level (Mill 1991, pp. 16–17).

In this chapter, we examine the state of Korea's democracy along three dimensions—attitudinal, behavioral, and structural—and find the Sewol disaster exposed serious problems in all of them (Barracca 2004).

We argue that one of the most powerful impediments to Korea's democratic consolidation is the political institutions that allow the president to exercise unbridled authority, and suggests that a proportional representation system is an effective institutional remedy for combating authoritarian elements in Korea's democracy.

Attitudinal Dimension of Democratic Consolidation

President Park's reactions to the criticisms of how she dealt with the Sewol disaster give us a glimpse into her attitude toward democratic principles. Democratic governments should guarantee the free flow of information and a high level of transparency regarding its policy-making processes. This is because a democratic government is, or at least should be, the agent of the people, and yet it frequently fails to live up to the principle. Therefore, a level of compromise on this principle could signal the government's commitment to democratic values.

During the crisis of the Sewol, Park largely remained silent and her government was extremely reluctant to share information with the public, betraying the principle of transparency and accountability. Her silence began to puzzle the public from the day of incident. The incident occurred at about 9:00 am and the president received reports at 10 and 10:15. Around 5 pm she visited the emergency control center to receive an oral briefing from the officials, which was broadcast on TV. In this afternoon meeting, she asked why it was so difficult to find and save the students when they were all wearing lifejackets. The question shocked the nation because everyone—except Ms Park—knew the reason: by then all the remaining passengers were trapped inside the ferry that had turned upside down. It was equally troubling for the public to learn that the president was missing between around 10 am and 5 pm. Even Kim Gichoon, her Chief of Staff, bluntly said that he did not know where she was (Choi 2014). As yet, no information about her whereabouts is available.

Park's reluctance to make public comments regarding the disaster was yet another indication of the government's reluctance to be transparent. The public had to wait for two weeks until April 29 to hear a much anticipated apology from the government, and it was quite disappointing. First of all, it came too late by Korean standards. In the past, presidents took much less time to deliver an apology and take responsibility in similar events. For example, Roh Myoohyun apologized three days after the subway fire in Daegu in 2003 that killed and injured 340 people, while Kim Youngsam offered his apology the very next day after the sinking of a ferry that killed 292 in 1993 (Kang S 2014). Second, the format of Park's apology was problematic. Instead of delivering it in front of the victims and their families, near the location of the incident, or in front of the press, she included an apology in her opening statement at a cabinet meeting, where it was buried in many other remarks. Third, it was short and disheartening. Although the word "apology" was included, the statement was anything but apologetic. She continued with her instructions to ministers regarding the aftermaths of the incident, which made her apology sound hollow. The opposition parties and victims' families reacted with anger (Kim Y 2014) and an overwhelming majority—63%—found the apology extremely unsatisfactory (Ji 2014).

The Sewol incident also revealed the Park administration's authoritative view on presidential authority. It acted as if the principle that its authority depends on the people seemed to disappear—even when the president became a target of the public's frustration and criticism—to the extent that it almost consecrated her as the republic's monarch. First, the government refused to recognize the presidential office's responsibility. Park's subordinates launched a political campaign to defend her reputation from a long string of criticisms of her handling of the crisis. Kim Gichoon voiced an opinion, which was repeated by other deputies, that Park's office was not legally in charge of crisis and disaster response. Therefore, he implied that criticism toward her and her office was not appropriate. According to him, the legal responsibility lay with the Central Disaster and Safety Countermeasures Headquarters (Kang and Park 2014). When Park finally responded to the criticism, she was busy attacking her critics. During a cabinet meeting in September, for example, she blamed a congressman who raised the question of her whereabouts on the day of the disaster and did not hide her indignation at such a question. Park acted as if to criticize her was to condemn the Korean nation and hurt the country—especially future generations (Joong-ang Daily 2014).

Park's authoritarian attitude was echoed in the actions of her deputies, who also prioritized their convenience above citizens' needs. On the day of the incident, when Suh Nam-soo, the Minister of Education, visited a temporary housing facility for victims' families, for example, he took a break to eat ramen using a table that had been used for medical aid (Kang S 2014). This scene appalled the public, as well as the emotionally and physically exhausted families, who had to tolerate it while sitting on the cold floor. On April 20, Song Young-chul, a high-ranking officer of the Ministry of Security and Public Administration, took a picture of him and his team in front of the list of victims, reflecting the popular selfies culture (Choi et al. 2014). This also infuriated the public because it showed the administration taking the incident in a light-hearted manner, as if its officials were going on a picnic. Although these events could be interpreted as isolated incidents or personal mistakes, they collectively betrayed the administration's attitude that it stands above the public.

The Sewol crisis exposed another disturbing pattern: the government's attitude shifted away from the democratic principle of transparency. The government manipulated the mass media in a poorly hidden but still effective way to mislead the public to believe that governmental efforts were actually effective. One infamous example came from the president herself. Toward the end of her speech that was televised live on TV on May 19, she began to shed tears, an act that did not look natural since Park continued her speech without wiping her tears. An analysis of her speech (Park G 2014) found an interesting pattern. The interval between eye blinks was 3.79 seconds on average, but towards the end of the speech, she stared at the camera without blinking for 31.9 seconds as tears began to flow down. The strangeness of this event was compounded not only by her lack of any reaction to the tears but also by a camera maneuver that slowly zoomed into her face. Her tears were widely considered a TV stunt, made together with the compliant media, to convince the public that the president genuinely sympathized with the grieving people.

The show of tears was not her first attempt to manipulate public perception. After she offered her first public apology during a cabinet meeting on April 29, President Park visited a memorial hall in the city of Ansan, where all the high-school student victims came from. The televised visit produced a famous scene in which Park hugged and consoled a sad-looking elderly woman, who most viewers thought was a victim's relative. However, this was yet another manipulated scene (Lee J 2014). In reality, when Park walked into the site, victims' families shouted at her out of frustration and anger and the security team quickly pushed them away from her. Afterwards, the president was guided by her aides to the elderly woman, who had apparently been instructed to follow the president at a close distance in an otherwise deserted memorial hall. It was later revealed that the person was not related to any of the victims, but recruited on site right before the president's visit. The public image that the president visited the memorial site, was welcomed by the grieving families, and gave them some condolence had thus been created by an orchestrated and untruthful manipulation of the media coverage.

The manipulation of public perception was not limited to a few staged political events. The rescue operation of the Sewol itself was a major target of the government's manipulation. Immediately after the ferry capsized, the government hurried to create the impression of a large and orderly rescue operation. On the day of the incident, Kim Soohyun, the head of the Western Coast Guards, announced to the public that the rescue was under way, utilizing 178 rescuers, 164 vessels, and 24 aircraft. Two days later, the government announced that 555 rescuers were working with 121 helicopters and 69 vessels. The information was widely and repeatedly reported in the mass media. However, the actual rescue scene on the sea was quite different. When the victims' families arrived at the site at the night of the incident, they were shocked to observe the dearth of rescue activities. The sea was unbelievably calm, and yet there was no ongoing rescue operation or even a rescue boat other than patrol boats circling around the Sewol. The absence of rescue operations notwithstanding, the government and the media repeated for the next few critical days the same story that large and active operations were under way (Choi et al. 2014; Newstapa 2014a).

The military was no exception. When reporters asked the Navy for more vivid pictures of its rescue operation, one officer was ordered to put on a diving suit during a media interview (Lee S 2014). Still feeling that it was inadequate, the Navy ordered others to spray water on him. This stunt was to mislead the viewers at home to believe that they were watching a hard-working rescuer who came back from the sea. Furthermore, when he told the reporters that the water condition was relatively calm, one of his superiors quickly intervened to have him reverse the comment so that their rescue efforts might look more heroic.

As shown in the aftermath of the capsizing, Park administration's efforts to mislead the public were as troubling as its incompetence. Although incompetence may be part of a large government, democratic or otherwise, a democracy is supposed to be better in terms of transparency in and out of government. Transparency enables democratic institutions and civil society to address the incompetence of political elites and government. Also, it provides a foundation on which informed citizens hold elected leaders accountable for their inadequacy through free and fair elections. This is why the political elite's attempts to manipulate public perception and knowledge are so detrimental to democracy, and the Sewol crisis unfortunately showed both the incompetent and the undemocratic faces of the Park government.

Behavioral Dimension of Democratic Consolidation

The Park administration's reactions to the Sewol sinking were alarming because of the degree to which the president's decisions and policies compromised the principles of democracy. First of all, from a constitutional perspective, her intrusions infringed on the principle of separation of power. Second, regarding the relationship between the state and civil society, her regression weakened civic liberty. Third, concerning human rights, the administration repeatedly violated basic human rights of Korean citizens. She was not unique: many democratic leaders engaged in undemocratic behavior and policies during times of national crises. President Lincoln suspended the principle of *habeas corpus*—the idea that one can protest an unlawful imprisonment before a court—during the Civil War without congressional approval in order to secure the capital; President Bush authorized the torture of detainees from the wars in Iraq and Afghanistan; and President Obama allowed indiscriminate espionage of the US public. However justifiable or not, such instances typically occurred in the presence of a threat to national security. In the case of the Sewol, no such threat existed, unless one would consider the powerless parents who had just lost their children and sympathetic citizens as a threat. Against them, Park's government behaved as if it were battling national security threats.

President Park threatened the basic principle of democracy by interfering in the legislative process of drafting a special law and a Congressional investigation into the crisis. Although Park herself proposed a special law in front of the victims' families on May 16, 2014, her administration later tried to weaken the power of the commission that would be created by the law to investigate the disaster. The families demanded that the commission be given the authority to subpoena witnesses and indict those responsible because they witnessed Congressional investigations degenerating into endless debates and foot dragging (People's Committee for the Sewol Ferry Tragedy). The public supported the families' demand: in July, a Gallup poll showed 58% of respondents supported the creation of a special committee with such powers (Jung H 2014). However, repeated negotiations among legislators remained unsuccessful (Kwak and Suh 2014).

Then, President Park intervened. In a cabinet meeting, she provided a legislative guideline. Park suggested that a commission as powerful as the victims' families demanded was not acceptable because they would violate the principle of separation of powers by compromising the authority of the prosecutor's office and ultimately threaten the legal foundation of the country. Park asserted that liberal democracy depends on an independent Congress, and the families' demand threatened this idea. Her instruction promptly hardened the Saenuri Party leaders' position and invigorated conservative newspapers to accuse the families of holding the country hostage.

Public opinion also shifted in favor of a political compromise. The final bill that passed the floor of the National Assembly and Park's desk in November largely reflected what Park wanted. According to the "Sewol Special Act," families can recommend only three members of a 17-member committee with power to summon witnesses but no authority to press charges (Kyunghyang 2014). Unfortunately for the families, this was not the end of their frustrations—the committee was further weakened by an enforcement ordinance attached to this law in March 2015. Known as the Presidential Decree for Sewol Special Act, it dramatically enlarged the discretional power of the government over the Parliamentary Special Investigative Committee and noticeably diminished the scope of the investigation (Park 2015). This obviously diluted what the public originally wanted and expected from the special law and seriously compromised the committee's autonomy.

Park's intervention practically aborted political negotiations between the major political parties and the victims' families and as a consequence, the prospect of a successful investigation—especially an investigation into the government's role and responsibility in the accident—became even more remote than ever. The result was, as expected, an investigation with no tangible results, frustrating the public as well as the victims' families, who were desperately waiting for accountability. It was clearly a case where the very idea that democracy should reflect the will of the people was overturned by Park's political calculation. As victim families' and public's frustration deepened, Park's undemocratic behavior became more apparent.

Censorship was another problem in the behavioral dimension of democratic consolidation. The Sewol tragedy was not only sad but also revealing of the shadowy side of South Korea's economic development, such as corruption in the safety and regulatory authorities (See Chap. 5) and the incapacity of authorities including the Coast Guard and the Blue House. The press had a number of opportunities to examine and report them, which should have helped citizens make informed decisions regarding their elected officials and the government. But Park's regime suppressed the freedom of the press in multiple ways, and sank the core idea of democracy deep into the mud (see Chap. 6). A leaked document revealed that the Korea Communications Commission (KCC)—the governmental body in charge of the press and mass communication—created a team to monitor and "control" press reporting of the Sewol disaster (Park J 2014). Moreover, Kim Sigon, a top official of Korean Broadcasting System (KBS), the national public broadcaster, and its labor union announced that its coverage of the Sewol disaster was filtered by its managers in order to minimize reports about the government's mishandling and responsibility. They argued that the root cause of such censorship was close ties between the president of KBS, Kilh Young-hwan, and the Park administration. Kilh, a political appointee of previous president Lee Myung-bak, aggressively intervened in the company's coverage of the Sewol incident, despite the law prohibiting such interference (Jung 2014; Lee B 2014). Mr Kilh himself confessed that the Blue House strongly pressured him to steer its news reports away from the government's mishandling of the Sewol disaster. According to him, the Blue House demanded that criticism be discouraged and critics dismissed, which left him no choice but to intervene (Kim and Lee 2014).

The Park administration's manipulation of the mass media was largely successful. Top managers of media companies seemed to have accommodated the wishes of the Blue House. For example, on May 5, Kilh called a meeting of managers and ordered them to stop criticizing the Coast Guard, and a news report about the Guard was hastily removed on the same day. Also, he tried to remove a caption from TV coverage reporting an opposition leader's criticism of the government, and manipulated the order of news items in the main news program, calling on the newsroom to move up coverage of the president to the second item. Such efforts resulted in biased coverage of the disaster and the government's reactions. All three major TV stations, KBS, the Munhwa Broadcasting Corporation (MBC) and the Seoul Broadcasting System (SBS), repeatedly reported misinformation, such as the number of passengers and the total rescued, and sent sanitized images of the president and her government. They all reported about the government's rescue operation using large resources such as "23 navy ships, 12 military aircrafts, and 1000 troops," (Kim U 2014) when in fact no rescue efforts were being made. KBS coverage of Park Geun-hye's April 17 visit to a stadium where the families of victims were staying was brazenly misleading. The scene was chaotic, filled with loud voices of concerned families. KBS, however, covered only those who welcomed her, completely editing out protesting families.

This practice was repeated during their coverage of Park's visit on May 4 (Kim H 2014). SBS suddenly canceled the production of an episode of an investigative program about the incident (Kim U 2014). TV coverage of the Congressional investigation was sketchy and scant too. From June 2 to August 30, a cable network, Joong-ang Tongyang Broadcasting Corporation (JTBC), broadcasted 22 reports about it while SBS had 10, KBS, 8.5 and MBC, 4 (Kim U 2014). The pattern of TV coverage was clear even though not uniform: inflation of the Park administration's efforts and deflation of its failure.

In addition to pressuring the media's management, the Park administration harassed the press with legal tools. The most frequently used one was defamation charges against those who were considered unfavorable to or critical of the government. The Blue House filed defamation suits against a liberal newspaper, the Hankyoreh (Kim S 2014), and the father of a victim (Kang M 2015). It even accused Tatsuya Kato, then Seoul bureau chief of Japan's Sankei Shimbun, on October 8, 2014. Mr Kato was indicted for his column quoting rumors that had originally appeared in the South Korean media about Park's mysterious disappearance on the day of the Sewol incident. In the column, Kato reported on the rumor that Park was dating a male partner during the critical hours, neglecting her duty. The insinuation was difficult to miss: the tragedy could have been lessened had she focused on the incident, and it certainly angered the Blue House. In reaction, the authorities barred him from leaving the country until April 2015 (Hong J 2015) and put him on trial. It was certainly unprecedented to put a foreign correspondent on trial and thus served as a strong warning to domestic reporters who did not have similar protection from the international community.

The Park administration's repression of civilian critics was much worse than its treatment of the media. Even online comment was not forgiven. In the southern city of Dae-Gu, a schoolteacher accused the President in a Facebook post right after the Sewol's sinking, and the reaction from the authorities was harsh and swift. The Board of Education of the city opened an investigation to determine if the teacher had violated her duty as a public employee to maintain political neutrality, and eventually disciplined her with an administrative measure (Yoon 2014). Artistic caricatures were not tolerated either. Artist Lee Ha was arrested for producing a caricature of the president that depicted her surrounded by dogs with the sinking Sewol in the background. One could read it as an accusation of Park, who was blinded by her loyal deputies and eventually missed the opportunity to save the ferry. The police first detained a volunteer who pasted copies on walls in May 2014 (Song 2014) and charged the person of violation of laws regarding outdoor advertisement. Later, Mr Lee himself was arrested in October when he threw thousands of copies of another caricature of Park from the top of a building in Seoul. This time, the charge against him was trespassing, even though the owner of the building did not press the charge (Joh H 2014a, b). In June, another citizen was arrested in Chun-cheon for drawing a picture accusing the president with the caption, "Decimator of the Sewol" (Park H 2014). Even a formal art show was censored. A giant poster of the Gwangju Biennale of 2014 became controversial when it included a painting of a chicken—an object often used to mock President Park—controlled by Kim Gi-chun, Presidential Chief of Staff, and Park Chung-Hee, President Park's father and former dictator of the country. The painting's message was obvious: President Park Geunhye was a puppet of her father and the old guard. The City of Gwangju announced the poster was inappropriate and would not allow the painting to be displayed at the exhibition, reflecting the political atmosphere of the country, where criticizing the president became a risky activity.

Even in dealing with the grievances of the victims' families and the public, the Park administration exploited police power as if they were a threat to her regime. On multiple occasions, the families noticed the presence of the police spying on them. For example, on April 26, the police was secretly recording the families' interview with the SBS (Chosun 2014), and later on May 17, the police clandestinely followed the families' entourage returning to Jin-do after their meeting with the president at the Blue House (Kim-yoon 2014). These were not isolated and rare incidents, but part of a large pattern. According to the police's own report to a congressman, the police actually deployed more than one thousand—accumulated figure—of its personnel in order to monitor the families in the city of Ansan, where the most student victims were from, and the main memorial park at Jin-do (Park, H.D. 2014). It is very likely that a much larger number of policemen were secretly monitoring them nationwide. Furthermore, the police did not just monitor the families but mobilized its personnel and equipment to prevent their movements. When the families began to march toward the Blue House three days after the incident as their patience with the government dissipated, they were quickly stopped by the police, who did not allow them to even leave the island of Jin-do, hundreds of miles from the destination (Gu 2014). This was but the beginning of the police's systematic control that recurred in the families' march to Yoido, where KBS is located, their parade to the Blue House in May (Lee C 2014), and many others.

Police repression of the public substantially increased in the first half of 2014, in which protests were often about the Sewol (Kim and Joh 2014). According to governmental records, the police mobilized 1.7 million officers for the first eight months of 2014, surpassing 2012's yearly total of 1.68 million. The increase in police mobilization is perplexing in light of the fact there was no significant increase in protests. According to the police, the number of protests increased only slightly from 9738 in 2013 to 10,504 in 2014. Nor was there any escalation of violence. The number of "illegal and violent protests" actually decreased from 51 to 45, as did the number of police injuries—92 to 78—during the same period (Statistics Korea). Thus, it is obvious that no upsurge of protests or violence preceded the larger presence of the police forces. The increase appears to be a result of political calculation. In other words, the police mobilization might have been the Park administration's reaction to the progressively grave political crisis—largely because of the Sewol crisis—because it had to turn to physical force to quell dissenting citizens.

Structural Dimensions of Democratic Consolidation

The structural base of Korea's democracy is riddled with potential defects, and the Sewol incident clearly demonstrated them. The high concentration of presidential power is the most important defect of all. The South Korean constitution allows the electorate to separately elect the president and lawmakers as in typical presidential democracies. However, unlike the US Constitution that encourages the balance of power between the president and Congress, the Korean constitution deters such balance by installing the post of prime minister between them. The prime minster serves the president, although congressional approval is necessary for the appointment. In this sense, the governmental structure is similar to the French system, known as a semi-presidential system. As Charles de Gaulle wanted and enjoyed, presidential authority is often more powerful in semi-presidential government than in other types of democratic regimes. Semi-presidential systems favor the dominance of the executive branch by far, and such dominance is usually deeply rooted in historical contexts and materialized in constitutions. Historically, semi-presidential regimes were usually born from the demand for a stable, and often dominant, executive branch. In France, de Gaulle revived the constitutional tradition of a strong monarch (Safran 1991: 11–15) as "the antithesis of its immediate predecessor, the Fourth Republic" that had 25 governments in 12 years (Elgie 1999: 75). Poland set up a powerful presidency to calm the heat of the country's speedy transformation by balancing the power of parliament, which was expected to be dominated by Solidarity (Krok-Paszkowska 1999: 173–177). In South Korea, early efforts to set up a parliamentary system failed, thanks to the political ambition of Syngman Rhee, who strongly advocated the presence of a monarchic figure in the government, the position he eventually took. His aspiration for overwhelming power and his victorious power struggle against the Assembly, after a massive arrest of Assemblymen, led to constitutional amendments that changed the type of government to a semi-presidential system in 1952 (Henderson 1988: 26–28). In general, semi-presidential regimes originate from historical contexts of either an ambitious, authoritarian leader (France) or an overly powerful parliament that needed to be checked (Poland), or both (South Korea).

Such historical backgrounds of semi-presidentialism are well reflected in the constitutions of countries with this system. First of all, in a semi-presidential regime, the legislature is typically weak in the legislative dimension, especially when a bill concerns financial contents or burdens. Most obviously the French parliament has a hard time initiating bills because bills and amendments can be blocked if they are considered to bring financial burden on the public (Article 40), and parliament cannot consider tax legislation without the consent of the executive. Likewise, in South Korea, the constitution does not allow the Assembly to modify tax bills without the executive branch's consent (Article 57). It is often the case that the government can refuse to consider a bill unless the legislature follows certain procedures. In France, for example, the constitution gives the government the power to object to "the consideration of any amendment which has not previously been referred to a committee" (Article 44). Article 15 of Poland's constitution put a similar restriction on legislating. Presidential veto is another barrier. In Poland and South Korea, the legislature needs the vote of two-thirds or more to overcome a presidential veto (Article 18 and Article 53, respectively).

Second, the legislature in a semi-presidential regime has weak authority against the executive branch. In a parliamentary system, the legislature can censure the government at will. The legislature of a semi-presidential regime does not have such complete and sole authority. In France, the National Assembly can pass a motion of censure against the government (Article 49); however, virtually no means to censure the president exists. Even where such means exist, it does not completely empower the legislature. In South Korea, the Assembly may impeach the president for the violation of a law (Article 65), but its censuring power is limited because only serious legal violations can lead to the removal of the president, and the Constitutional Court (Article 111), not the Assembly, makes the final decision regarding the fate of the president. The prime minister of South Korea is subject to the Assembly's recommendation for removal (Article 63), but it is only a recommendation. In addition, as often seen in South Korea, the post of prime minister serves as a major source of the president's institutional power. When pressure mounts, political troubles are often passed to the prime minister, who is obligated to shield the president and often takes responsibility. The president can dismiss the prime minister as a last resort and such a change of the face of the government—especially with an overhaul of cabinet members—often alleviates or removes altogether political pressure on the president. Presidents of a more traditional presidential system as in the USA do not have this powerful political tool. Neither do prime ministers of parliamentary democracies. Of course, this option is possible when the president can control the nomination of the prime minister—in other words, a majority in parliament.

Compared to the weak legislature, the presidential authorities of semi-presidential systems are powerful and unrivaled. As long as the president enjoys the majority support of the legislature, the president can set the legislative agenda for the legislature, take political initiatives at will, and control national discourse, as well as specific policies. In the case of South Korea, the president has the authority to declare war (Article 73); to issue presidential decrees (Article 75); to use emergency powers (Article 76); and to proclaim martial law (Article 77). Now compare these: the US president has no authority to declare war or proclaim martial law, while the British prime minister lacks the power to govern the country via executive orders. All things considered, the balance of the power is heavily tilted, favoring the president over the legislature. In other words, structure-wise, the South Korean president is well poised to act like a monarch at any given political opportunity. Therefore, it should not be surprising that Park took the opportunity to do so. During the Sewol crisis, Park Geun-hye remained unyielding to the victims' families' and the public's demand for more active rescue operations for the victims. Even more troubling, she refused to even communicate with the public over the crisis. She did not engage the public, except for a few carefully staged occasions. Instead, Park stubbornly isolated herself within the high walls of the Blue House. Her form of public communication was largely limited to giving out orders during cabinet meetings. They were then relayed to the public through the media or spokespersons. Beyond this, not much else was shared with the public. Such secrecy by Park troubled many who were anxious to learn more about the situation and the government's response from the head of government. However, virtually no one from the government and the governing Saenuri party raised the issue of secrecy. No political actor or tool existed to lessen the public's frustration and disappointment. Many factors caused the helplessness of the public, but Park's unparalleled power was undoubtedly crucial. Especially with the second term barred by the constitution, the president fears no one, not even the voters. Should the president choose to be authoritative, as Park did, the South Korean institution cannot do much to stop it, especially with her party in firm control of the National Assembly. This is a major problem for South Korean democratic consolidation, and it can deteriorate further in the future.

Closing Remarks

The Sewol incident has revealed the shortcomings in Korea's democracy. Attitude-wise, the political elites acted as if they were the feudal masters of the citizens, not their servants. Especially regarding the president, their attitude deviated from the general norms of a modern, developed, and democratic republic. Behavior-wise, the government continued to be repressive toward its citizens. As a result, even the most basic human rights and civil liberties have been compromised and threatened. Also there are structural deficiencies that complicate Korea's democratic consolidation. One of the main structural problems is the high concentration of political authority in the hands of the president, allowing her to act like a monarch.

While all the three dimensions must be addressed in order to consolidate democracy in Korea, I would suggest the constitution be amended in order to strengthen the legislature in the short run and eventually create a parliamentary system. Even Barracca pointed out that progress in a behavioral dimension is the most critical element of democratic consolidation—it will come and be sustained only after structural conditions are met and are ripe. My advocacy for a stronger legislature is based on the fact that it is a pillar of democracy as a bridge between the governing and the governed, which has been echoed over and over again in academia. Combined with the legislature's influence over the government, public access to the government through the legislature serves as a critical avenue for the public to the political authorities.

I have no illusion that the proposed structural fixes would be the panacea for the challenges witnessed during the Sewol crisis. However, it will certainly limit the powers of the head of state and create conditions that could prevent him or her from acting like a monarch, as Park Guen-hye did during the Sewol disaster. Park might have had no choice but to immediately and forcefully redirect her attention toward the incident during the early hours if she had been faced with the prospect of another election, such as a US presidency or the European-style prime ministership. In a similar vein, Park could have been forced to be less secretive and more open with the public. Major mistakes and misbehavior of leaders often change the political calculus of subordinates and have the potential to unsettle the political horizon. Even members of the same party might and could withdraw their loyalty to the leadership in cases of plunging popularity, as was the case when the Labour Party ousted former British Prime Minister Tony Blair in 2007. Thus, it is entirely possible a different political system—such as a parliamentary government—could have forced Park to act more responsibly or even to resign. Members of the Saenuri Party might have confronted Park as well, worrying about their prospects for re-election. As for dealing with the victims' families and leading the investigation of the sinking, the Park administration might have chosen a different path if she had had a strong incentive to listen to the public and to compromise with other politicians. That incentive, of course, could be given when there is a looming election or the need to maintain multiparty coalitions to govern. Therefore, structural changes to South Korean democracy are by no means a magic potion, but nonetheless a necessary step for better governance and democratic consolidation of the country.

References

Barracca, S. (2004). Is Mexican democracy consolidated? *Third World Quarterly, 25*, 1469–1485.

Bulbeobpokryeoksiwi mich gyeongchalgwan busangja balsaeng [Illegal and violent protests and injured policemen]. *Statistics Korea*. http://www.index.go.kr/potal/main/EachDtlPageDetail.do?idx_cd=1613. Retrieved from August 23, 2015.

Choi, J. J. (2014). Sewolho Sago Dangil, Park Geunhye Daetongnyeong Haengbang Myoyeon [On the day of sewolho sinking, president park's whereabout is unknown]. *Ohmynews*.

Choi, M., Lee, G., Song, Y., & Kim, H. (2014, May 7). Sewolho Chimmoreseo Chamsa Kiun Busil Daeeungkkaji [From sewolho's sink to tragic reactions]. *The Kyunghyuang Shinmun*.

Dahl, R. (1989). *Democracy and its critics*. New Haven: Yale University Press.

Eongteo-ri Seonbakgeomsatpsyeonbong Chaenggigi...Munje Danche Hanguk-seongeup ['Corruption in inspection of ships and embezzlement ... troubled Korean register']. *JTBC News at 9* (April 23, 2014).

Geugeonni Algosipda' Sewolho Chimmol Daruneun Jejakjin Inteobyu Mollae Nogeumhan Sabokgyeongchal ['Is the policemen in plain cloth who secretly recorded the staffs of the tv show 'I Want to Know?']. *The Chosun Ilbo* (April 27, 2014).

GNI per capita, PPP. World Bank. http://data.worldbank.org/indicator/NY.GNP.PCAP.PP.CD. Retrieved from August 10, 2015.

Gu, S. (2014, April 20). U-ri Aireul Sallyeonaera Kkeunnae Cheongwadae Mot Gan Siljongja Gajokdeul ["Save My Child" Victims' families failed to reach the blue house]. *The Huffington Post Korea*.

Hong, J. (2015, April 14). Gato Jeon Sankei Jigukjang Chulguk [Sankei reporter leaves Korea]. *Kyunghyang Shinmun*.

International Trade Statistics 2014. World trade Organization. https://www.wto.org/English/res_e/statis_e/its2014_e/its2014_e.pdf. Retrieved from August 10, 2015.

Ji, Y. (2014, May 2). Sewolho Minsim Bakgeunhyeege Deung Dollin Du Gaji Iyu [Why did public turned against park on sewolho']. *Ohmynews*.

Joh, H. (2014a, October 22). Park Geunhye Myeongyehweson, Mudeogigobal Susa [Multiple investigation of defamation of park]. *Media Today*.

Joh, H. (2014b, October 20). Park Daetongnyeong Pungja Jeondan Ppurin Pop-art Jak-ga Yeonhaeng [Pop artist arrested for his art piece mocking park]. *The Kyunghyang Shinmun*.

Jung, C. (2014, April 24). Lee Jeonghyeon Hongbosuseok, Gijadeurege "Han Beon Dowajuso" [Lee, Chief officer of blue house asked reporters to help]. *Media Today*.

Kang, M. (2015, March 17). Kim Jaewon, Yeeunappa Yu Gyeonggeun Ssi Myeongyehweson Goso Nollan [Kim Jaewon Sue Yu Gyeonggeun—Father of Yeeun—and Cause Controversies]. *Ohmynews.*

Kang, S. (2014a, April 19). Palgeori Uijaeseo Ramyeonmeokgo… Janggwannim, Yeogi Wae Osyeonnayo' [Sitting comfortable to East Ramen … why did minister come?]. *Ohmynews.*

Kang, (2014b, April 29). Sagwae Insaekan Bakgeunhye Daetongnyeongttgwageo Daetongnyeongdeureun? [President Park Not Liking Apologies … How About Her Predecessors?]. *Kukmin TV.*

Kang, G., & Park, G. (2014, July 10). Kim Gichun Cheong Jaenankeonteuroltawo Anya … Haegyeong Jeonmunseong Bujok [Kim argues the blue house is not a control tower and the coast guards lacks specialties]. *Yoenhap News.*

Kim, H. (2014, May 9). Dodaeche EottaetgillaettKBS Sewolho Bo-do 7Gaji Munjejeom [Seven problems of KBS coverage of sewol]. *The Hankyoreh.*

Kim, H., & Lee, J. (2014, May 16). Sajangi Ulmyeo Daetongnyeong Tteunnirago Satoe Jongyong [Crying KBS president asked to resign because this was the president's wish]. *The Hankyoreh.*

Kim, J. (1993). Korea's 14th presidential election and economy. *Korean Political Science Association, 27,* 99–120.

Kim, J., & Joh, H. (2014, October 23). Neureonan Gyeongchal, Jeonggwonui Howimusa Dwaetda [Increased police force has become the regime's bodyguard]. *Sisa Journal.*

Kim, S. (2014, December 24). Cheongwadae, Sewolho Gwallyeon Hangyeore Sangdae Sosong Paeso [The blue house lost in court against The Hankyoreh]. *The Hankyoreh.*

Kim, U. (2014, December 19). Sewolho Giregineun Eotteoke Mandeureojige Dwaenna [How reporters become so useless']. *The Pressian.*

Kim, Y. (2014, April 29). Park Geunhye Sewolho Sagwae "Igeon Sagwa Anida" [Park's apology was not accepted]. *Media Today.*

Kim-yoon, N. (2014, May 20). Gyeongchal, Sewolho yujok mihaeng pamun … Yujok "bulbeob sachal [Police Spied on Sewol Families … Families accuse illegal]. *The Pressian.*

Kwak, J., & Suh, U. (2014, August 7). Sewolho Yujokdeul Park Yeongseoni Dwitongsu …Yeoya Habui Burinjeong [Sewol Families Blame Park Yeongseon stabbing their back … Don't Support for the deal]. *Pressian.*

Lee, B. (2014, May 18). Gil sajang 8ildongan 3charye bodo gaeib [Mr. Gil Intervened 3 times in 8 days]. *Kyung-hyang Shinmun.*

Lee, C. (2014, May 14). Sewolho yugajok, eonron baro jabeureo KBSeseo cheongwadaekkaji [Sewol Families march to KBS and the Blue House to correct media coverage]. *Media Today.*

Lee, J. (2014, May 1). Hwaksandoeneun bunhyangso halmeoni yeonchul uihok [Growing controversy over elderly woman's visit to memorial site]. *Media Today*.

Lee, S. (2014, April 24). Haegunui bangsongyong geurimmandeulgi [Navy's manipulation for the media coverage]. *Sports Hankook*.

Lijphart, A. (1993). Constitutional choices for new democracies. In L. Diamond, & M. Plattner (Eds.). *The global resurgence of democracy*. Baltimore, MD: Johns Hopkins University Press.

Locke, J. (1924). *Of civil government: Two treatises*. London: J.M. Dent.

Mill, J. (1991). *On Liberty and Other essays*. Oxford and New York: Oxford University Press.

Park, G. (2014). Park Geunhye nunmul bunseok [Analyzing park's tears]. *Youtube*. Retrieved January 28, 2016, from https://www.youtube.com/watch?v=Er2WBXDS-G8.

Park, H. (2014a, April 10). Jeong Gheongrae "Sewolho yugajok sachale 1000myeong neomneun gyeongchal dongwon" [Congressman Jung, More than 1000 policemen spied on victims' families]. *The Kyunghyang Shinmun*.

Park, H. (2014b, June 1). Park Geunhye daetongryeong pungja geurim tto balgyeon...20dae hyeonhaengbeom geomgeo [Found another caricature of Park Guenhye ... Man was arrested at the scene]. *Newsis*.

Park, J. (2014, April 28). Park Geunhye jeongbu, sewolho 'bodotongje' mungeon mandeuleossda [Park administration prepared manual to control Sewol coverage]. *Media Today*.

Park, J. (2015, April 3). Sewolho teukbyeolbeob sihaengryeong, kkoriga momtong dwiheundeuna [Presidential decree for sewol special act, tail rattles the body]. *The Korea Times*.

Park Daetongnyeong, "Daetongnyeong Modok Bareon Do-ga Jinachyeo" Seolhun Uiwon Gyeonyang? ['President park, "Insulting Presidency Is Not Welcomed" is she talking to congressman sul?']. *The Joong-ang Daily* (September 16, 2014).

People's Solidarity for Participatory Democracy. (2014). We want the truth of the Sewol Ferry tragedy. Retrieved August 11, 2015, from http://www.peoplepower21.org/PSPD_press/1192232.

Sewolhoteukbyeolbeob deung 'Sewolho3beob', chamsa 205il mane gukhoe bonhoeui tonggwa ['Sewol Special Laws Pass 205 days after the sink']. *Kyunghyang Shinmun* (November 17, 2014).

Sewolho goldeuntaim, gukganeun eobseossda [Missing the golden time, no state existed]. *Newstapa* (July 24, 2014a). https://www.youtube.com/watch?v=Fm81fi5F5EE.

sewolho teukbyeolbeob, ohaewa jinsil [People's committee for the sewol ferry tragedy 'Sewolho special laws, misperception and facts']. http://sewolho416.org/2287. Retrieved from January 28, 2016.

Song, H. (2014, May 26). Daetongryeong pungja seutikeo butyeossdago gyeongchal jambok susa'kkaji [Police utilizes aggressive tactics to deal with caricature of Park]. *The Hankyoreh*.

Yoon, G. (2014, June 10). SNSro daetongryeong bipanhan joe?... Gyosa gyeonggo [Teacher is warned for online criticism of president]. *Ohmynews*.

Author Biography

Taehyun Nam is Associate Professor of the Department of Political Science and Director of International Studies at Salisbury University, Maryland, USA. His research focuses on mobilization and political development and he is the author of several books in the USA and the Republic of Korea.

CHAPTER 5

Capturing Collusion: The Industry and the Government in Ferry Safety Regulation

Jong-sung You and Youn Min Park

Introduction

The Sewol ferry accident was a man-made disaster. The passenger ship capsized because its crew and safety regulators violated basic safety regulations, and its capsizing led to the death of many passengers because the crew and rescuers failed to implement effective emergency measures and rescue operations. Moreover, the Sewol tragedy revealed that violations of safety rules were not uncommon but routine practice in the passenger shipping industry. Shippers knew that they did not have to comply with safety regulations, and regulators at the frontline knew that the system was not working. What caused such wilful negligence on the part of both the shippers and the regulators? This chapter addresses this question.

While some blamed corruption for safety regulatory failures (Hong 2014; Roh 2014), the prosecutorial investigations found little evidence of outright corruption such as an exchange of bribes and favors. Others

This chapter draws substantially from You and Park (2017).

J. You (✉) · Y.M. Park
Australian National University, Canberra, Australia

J.-J. Suh and M. Kim (eds.), *Challenges of Modernization and Governance in South Korea*, DOI 10.1007/978-981-10-4023-8_5

attributed safety regulatory failures to neoliberal reforms such as the extension of passenger ships' maximum life from 25 to 30 years and the delegation of authority for important safety regulation enforcement to the shipping industry association, or industry self-regulation (Ji 2014; Lee 2015; Oh 2014; Suh 2014; Woo 2014).[1]

But our research shows that the role of neoliberal reforms in causing the tragedy was rather limited, though not negligible. In this chapter, we instead develop the argument that the dismal safety regulatory failures were largely caused by regulatory capture, or "regulators being captured by the regulated industry" (You and Park 2017). Because the government and regulatory agencies were effectively captured by the ferry industry and Cheonghaejin Marine Co., the owner of the Sewol ferry, which only pursued profit without due consideration of passenger safety, they were unable to create effective regulations and enforce them rigorously. The government failed to impose international safety standards on the domestic passenger shipping industry, and instead continued to allow the shipping industry association to hire and control the vessel-operational managers, who were the frontline enforcers of safety regulations. It continued to protect monopolies in the ferry business while relaxing safety regulations such as the ship age limit, not because its regulation and deregulation were guided by economic logic but because its decisions were directly influenced by the industry lobby.

In addition, our process-tracing shows that the private delegation of safety regulatory authorities and some features of public–private partnership in rescue operations were not introduced as neoliberal reforms but as state-corporatist arrangements (You and Park 2017).[2] We find that since the Park Chung-hee regime first introduced the industry self-regulatory scheme as a state corporatist arrangement in 1972, the arrangement has persisted, leading to the proliferation of parachute

[1]The enforcement regulations for the Marine Transportation Act were revised in January 2009. Before the revision, passenger ships' age limit was 20, with possible extension of up to 5 years. The revised rule stipulated the same 20-year limit, but allowed extension of up to 10 years. Effectively, the maximum life for passengers ships was extended from 25 to 30 years.

[2]Process tracing, or causal process observation, is an effective tool to find empirical evidence for the causal processes (Bennet and Checkel 2015). Our process tracing is based on the examination of a variety of primary and secondary source materials including earlier and current versions of the laws and regulations governing marine safety and archival documents of the government and marine industry.

appointments and collusive relationships. What looked like the privatization of rescue operations, for example, was in fact a collusive scheme made between the Korean Coast Guard and salvage companies on the basis of the corporatist arrangement (You and Park 2017). The Sewol tragedy was an incident that resulted from the enduring legacies of the state-corporatist arrangements made during the Park Chung-hee period, a persistent feature of the authoritarian developmental state that has been largely neglected by the literature on Korea's political economy.[3]

Also, we find that regulatory capture was an important cause of the Fukushima nuclear power plant accident in 2011 (Howe and Oh 2013; National Diet of Japan 2012; Synolakis and Kânoglu 2015; Wang and Chen 2012; Yokoyama 2013). Although we do not go into the details of the Fukushima nuclear accident in this chapter, we provide a brief comparison between the two accidents focusing on the issue of regulatory capture.

In the next section, we offer a brief conceptual discussion of corruption and regulatory capture and then discuss state corporatism and neoliberalism in Korea's political economy. In the following section, we examine the evidence of regulatory capture that led to the Sewol ferry accident. Then, we scrutinize the process by which regulatory capture was institutionalized, focusing on the enduring legacies of state corporatism in Korea. We also compare the regulatory capture manifested in the Sewol tragedy with that in the Fukushima nuclear accident. The concluding section discusses policy implications.

[3]The large body of developmental state literature on Korea neglected Park Chung-hee's transformation of business associations according to the principles of state corporatism, focusing on the Park's *questionable* contribution to the establishment of a meritocratic bureaucracy (You, forthcoming). There have been few studies of corporatism in Korea. Robert Wade (1990), Ziya Onis (1991), and Jonathan Unger and Anita Chan (1995) gave brief descriptions of state corporatism of the developmental states in Korea and Taiwan, but their main focus was not Korea. Moon-Kyu Park (1987), Jang Jip Choi (1989), Yun-Tae Kim (1998), and Dennis McNamara (1999) are among the few studies of Korean corporatism written in English, while Young-Rae Kim (1987), SaKong and Kang (2001), and Eui-Young Kim (2005) are notable studies of Korea's interest groups and state corporatism written in Korean. Lee (2003: 104) briefly describes Park Chung-hee's state-corporatist organization of business associations.

Conceptual Discussion

1. Corruption and Regulatory Capture

Corruption and regulatory capture are often blamed for man-made disasters (Lewis 2011). Penny Green (2005) examines the role of corruption in magnifying the impact of earthquakes. An important study by Soliman and Cable (2011) analyzes the role of corruption in the 2006 sinking of an Egyptian ferry in the Red Sea, which killed 1034 people. They argue that high-level corruption not only caused the disaster but also exacerbated its impacts. In the aftermath of the Fukushima nuclear accident in 2011, the Fukushima Nuclear Accident Independent Investigation Commission concluded that regulatory capture was largely responsible for the man-made disaster (National Diet of Japan 2012).

Before we examine the role of corruption and regulatory capture in causing the Sewol accident, let us begin with conceptual clarifications. Corruption is typically defined as "misuse of public office (or entrusted power) for private gain," while capture is defined as the state or policy process "losing autonomy and serving powerful private interests away from the public interest" (You 2015: 8–9). As there are many types of corruption such as petty vs. grand corruption, and political, bureaucratic, and judicial corruption, so there are various types of capture such as state capture and regulatory capture. In a principal–agent–client relationship, in which the people are the ultimate principal and the elected and appointed public officials are the agent, corruption can be understood as the agent's misuse of power, while capture represents a situation in which the agent loses his autonomy and serves the client's interest rather than the principal's (You 2015: 23–30). Capture may involve corruption, but it can also occur through influence without outright corruption. Regulatory capture is defined as regulatory agency being captured by the regulated industry, but the practical meaning of this term has been understood very differently by scholars.

There are two contrasting views on regulation (Dal Bó 2006). The traditional public interest view posits that regulation is needed to prevent or reduce market failures and enhance public interest. In contrast, a "regulatory capture" theory put forward by Stigler (1971) challenges this view, arguing that "regulation is acquired by the industry and is designed and operated primarily for its benefit." These two views have led to the regulation–deregulation debate, with economic benefits and costs of regulation becoming the focus of this debate.

While the regulatory capture theory is often associated with the deregulation camp, it should be noted that regulatory capture can lead not only to the creation and maintenance of regulations favouring insiders such as an entry regulation but also to the removal or relaxation of socially beneficial regulations to reduce the costs of compliance. Indeed, from what we saw in the 2007–2009 global financial crisis, it was not excessive regulation but deregulation driven by regulatory capture that was responsible for the gross failures in the US financial system. Now, there is an increasing consensus that deregulation can also result from regulation capture by an industry (Carpenter and Moss 2014). Evidence of regulatory capture can be found in a regulatory policy shifting away from the public interest toward an industry interest, and in the action of the industry in pursuit of this policy shift being sufficiently effective (Carpenter 2014).

In the scholarly and journalistic discussions about the causes of the Sewol tragedy in Korea, the conceptual distinction between corruption and capture is often ignored. For example, Hong Sung-Tae (2014: 132) argues that the Sewol tragedy was not an *accidental* accident but a *normal* accident generated by the *corrupt* society of Korea. He notes that one of the striking similarities between the Namyoung Ferry accident in 1970, which claimed 338 lives, and the Sewol Ferry accident in 2014 is that corruption not only caused the sinking but contributed to the dismal rescue failure (Hong 2014: 138). However, his use of the term corruption is based on a broad definition that includes collusion and capture. In this chapter, we will heed the conceptual distinction between corruption and regulatory capture.

2. State Corporatism and Neoliberalism in Korea's Political Economy

While the existing literature on Korea's government–business relations is focused on the national level, it is important for the purpose of our chapter to examine the organization of interest mediation and regulatory arrangements at the industry level. In this regard, the concept of corporatism, as opposed to pluralism, is useful for understanding Korea's political economy. Although recent political and scholarly discourses in Korea turn their attention to neoliberalism, many of them conflate important differences in the sources of deregulation and privatization and turn a blind eye to a corporatist structure that exists between the state and businesses at the industry level. To adequately explain the presence of both regulation and deregulation in the passenger ferry industry in particular and Korea's industries in general, we need to pay close attention to the corporatist structure that organizes interest mediation among state and private actors in a particular way.

Philippe Schmitter (1979) defines corporatism as "a system of interest intermediation, in which a limited number of singular, compulsory, non-competitive, hierarchically ordered, and functionally differentiated interest groups are recognized by the state and granted a representational monopoly in exchange for certain controls," as opposed to pluralism, in which various interest groups compete for influence over policy domains. He distinguishes between state (or authoritarian) corporatism and societal (or democratic) corporatism, depending on whether the corporatist structure has been created, imposed or controlled top down by the state or formed bottom up autonomously. While societal corporatism is best exemplified by Sweden, Switzerland, the Netherlands, Norway, and Denmark, state corporatism is represented by Portugal, Spain, Brazil, Chile, Peru, Mexico, and Greece, as well as the pre-war experiences of Fascist Italy, National Socialist Germany and Austria under Dollfuss. Cawson (1986: 114) notes that corporatist arrangements can be made at the national (macro), sectoral (meso), or firm (micro) level. Meso-corporatism in the form of a negotiated industrial policy may involve protectionist measures in exchange for self-regulatory actions, often imposed by the state. A necessary precondition is the associational monopoly in the industrial sector.

Neoliberalism is generally understood as radical market fundamentalism that emphasizes liberalization, deregulation, and privatization (Young 2011). Liberalization is to allow the unfettered cross-border movements of capital, goods, and services; deregulation is to remove the government control of economic activities; and privatization transfers previously publicly provided services to the private sector. While classical liberalism advocated free markets in opposition to mercantilism, neoliberalism emerged in response to the expansion of welfare capitalism and the associated big government. In general, neoliberalism tends to be more closely associated with pluralism rather than corporatism.

Park Chung-hee (1961–1979) reorganized the state-society relations according to the principles of state corporatism (Wade 1990: 27,

294–295). Emulating pre-World War II Japanese state corporatism, the government recognized only one association for each industry as both a channel of monopoly interest representation and a tool of state control (Kim 1987, 1998; Lee 2003; Park 1987). The major role of the industry associations was to collect information and mobilize business-sector support for government policies (Park 1987). The government also delegated to the associations various authorities such as the registration of business, inspection, and assessment works, as well as supervision and sanctions over members. Thus, they often acted as a regulator on behalf of the government (Kim 1987, 2005; Lee 2003: 104). These arrangements fit the characteristics of meso-corporatism defined by Cawson (1986).

While the close relationship between the government and business was prone to rent-seeking and collusion, its nature has evolved over time. State control over business has weakened with economic liberalization, as well as political democratization since the 1980s. Despite the weakening of state control, the state's protection of the large chaebols and industry associations continues as the collusive relations become entrenched in each industry.

In the aftermath of the 1997 financial crisis, the Kim Dae-jung government (1998–2002) implemented a sweeping reform of the financial sector and corporate governance, complying with the International Monetary Fund (IMF)–mandated neoliberal reform agenda (Haggard 2000; Moon and Mo 2000). The Kim Dae-jung and subsequent governments also experimented with societal corporatism (or democratic corporatism) at the national level, notably through the Tripartite Commission, although this experiment was largely unsuccessful except for the first few months of its inception (Kong 2004).

There are different views about whether the post-crisis reforms have dismantled the developmental state in Korea. Some scholars argue that the Korean state has been radically transformed into a neoliberal state (Ji 2013; Kalinowski 2008; Lim and Jang 2006; Pirie 2012; Yeung 2014). Others argue that the developmental state is still alive or at least significant elements of it remain intact (Jang 2014; Kim 2012; Lim 2010; Park 2011; Stubbs 2009; Um et al. 2014). Our study is located in the context of the ongoing debate about the extent of the neoliberal transformation and the enduring legacies of the authoritarian developmental state in Korea.

Regulatory Capture in the Sewol Accident

Investigations have shown that violations of basic safety regulations such as overloading, poorly secured cargo, and inadequate ballast water directly contributed to the capsizing of the Sewol ferry. Vessel-operating managers in charge of safety regulations were supposed to inspect these conditions before the departure of vessels, but they only looked at the load line through a telescope from a distance without physically checking the vessels.[4] This procedure might have made the Sewol more unsafe rather than safer, for the Sewol crew removed more than half of the required ballast water in order to compensate for the overloading and make the load line visible above the sea surface, the only indicator the vessel-operational managers checked. This action severely impaired the stability of the vessel.

The vessel-operational managers were hired by the Korea Shipping Association (KSA), the interest group that is formed and financed by ship owners to represent their interest, thanks to a collusive arrangement made with the government in the early 1970s (see Sect. 4 below). It is thus not surprising that some vessel-operating managers who had attempted to impose strict safety rules were often rebuked and disadvantaged by their employer (Seo 2014). The industry self-regulatory system was inherently ineffective because frontline regulators were effectively captured by the regulates by institutional design.

Another failure was made in the process of inspecting the Sewol's structural modification. The Korean Register (KR), a private entity that had the legal authority to inspect and classify vessels, approved all features of the Sewol in spite of risky renovations and false reports submitted by the Cheonghaejin, the owner of the Sewol (BAI 2014). It is curious that the KR's inspection was so slack that later proved fatal, even though the prosecution found no evidence of bribery.

These failures were not isolated events. There was in fact a grave violation of regulation at every stage of the Sewol's operation, from getting approval for adding a new vessel and the renovation of the vessel to the creation of the vessel-operational management rules and the clearance for

---

[4] Load line is a special marking positioned amidships, which depicts the draft of the vessel and the maximum permitted limit in distinct types of waters to which the ship can be loaded.

departure from the port. Government agencies and private entities with the authority to inspect, approve, and monitor these processes all failed to fulfill their duties (You and Park 2017). The question is whether these failures were caused by simple human errors, some individual instances of corruption, or a systemic capture of the regulators by Cheonghaejin and the ferry industry. Our research shows that there was indeed a systemic regulatory capture. We present evidence of a capture in three broad areas: (1) collusive ties between the government and the industry formed by parachute appointments; (2) a failed legislative attempt to impose the international safety management (ISM) code and to reform the ineffective vessel-operational managers system; and (3) selective deregulations that were contrary to neoliberal rhetoric.

1. Government–Business Collusion Due to Endemic Parachute Appointments

The delegation of regulatory enforcement to private entities such as the Korea Shipping Association and the Korean Register should have been accompanied by strong and effective government oversight. But the Ministry of Oceans and Fisheries and the Korean Coast Guard failed to serve the public's interest by offering independent oversight, and instead had an incentive to work with the passenger shipping industry because of endemic practices of parachute appointments.

Many senior officials who retired from these regulatory agencies were employed by the non-profit organizations with the delegated regulatory authorities such as the KSA and the KR. As many as 10 out of the 12 people who have served as chairman of the KSA had previously held a senior position in a major government body, such as the Ministry of Oceans and Fisheries, that was supposed to oversee its performance. Also, of the 12 former chiefs of the KR, eight had previously worked for the ministry or ministry-affiliated organizations as senior officials (Lawyers 2014, 107). The media and the public called the collusive relationship between the ministry and the industry formed through these parachute appointments *Hae-fia* (sea mafia), and more generally *Gwan-fia* (bureaucratic mafia).

The literature on regulatory capture shows that the "revolving door" phenomenon, or frequent personnel movement between the government and business, can render regulatory agencies susceptible to capture by the very industry that they are supposed to regulate (Horiuchi and

Shimizu 2001; Makkai and Braithwaite 1992; Salant 1995). Ex-officials may lobby current officials for less rigorous oversight, and incumbent officials who expect a post-retirement career in the regulated industry may be lax in their regulatory enforcement and feel obliged to favor the industry at the expense of the public interest.

In South Korea, pervasive parachute appointments must have created a conflict of interests, leading to collusion and capture (Hong and Lim 2016). The problem was particularly serious between the marine safety regulatory agencies and the domestic passenger shipping industry.

2. Failed Legislative Attempts to Strengthen Safety Regulation

The South Korean government has worked to improve marine safety over the years. In particular, it attempted to introduce the ISM code to both the domestic and the international shipping industries. This attempt failed because of the industry lobby, which strongly suggests regulatory capture. There was also a failed legislative attempt to improve the ineffective vessel-operational managers system. Had either attempt been successful, the Sewol ferry accident might have been avoided.

The International Maritime Organization (IMO) created the ISM Code to improve maritime safety in 1994, recognizing that a majority of maritime accidents were man-made, for which an inadequate safety regulatory system was largely responsible. Korea's Ministry of Oceans and Fisheries announced plans in 1996 to introduce the ISM code to both international and domestic shipping. While the international shipping industry complied with the government plan, the domestic shipping industry resisted. The KSA demanded that the government postpone the planned implementation of the ISM code for domestic shipping. Accordingly, the effective date for domestic passenger shipping was deferred from July 1, 2001 in the Regulatory Reform Committee's initial draft to July 1, 2002 in its final draft and to January 1, 2003 in the government's bill (KSA 2012: 558–560).

After the Maritime Transportation Safety Act was amended in early 1999 to mandate the application of the ISM code to the ferry industry starting from 2003, the KSA changed its lobbying strategy. It argued that the existing vessel-operational managers system was more effective than the ISM code. The Regulatory Reform Committee accepted the industry's argument and decided to maintain the vessel-operational managers system and to exempt domestic passenger shipping from the ISM code. Accordingly, the Maritime Transportation Safety Act was amended again on December 26, 2002, just days before the ISM code was supposed to take effect for the domestic passenger shipping (KSA 2012: 560–563).

It is difficult to understand why the Ministry of Oceans and Fisheries and the Regulatory Reform Committee changed their policy in just a few years. While they had decided to replace the ineffective vessel-operational managers system run by the KSA with the new safety regulatory system based on the ISM code between 1996 and 1999, they reversed their decision in 2002 to maintain the existing system. This curious change of the government's regulatory policy is hard to explain without considering regulatory capture.

There was another opportunity to improve the ineffective vessel-operational managers system in 2011 when some National Assembly members proposed a bill to amend the Marine Transportation Industry Act to make vessel-operational managers independent from the Korean Shipping Association. However, the bill was killed off because of strong opposition from the Ministry of Oceans and Fisheries. The ministry's rationale was that the existing system was working fine and that a new system would require a substantial budgetary commitment (Seo 2014). Although we were unable to find any evidence of capture in this legislative episode, the ministry's opposition might well have represented the bureaucrats' common interest shared with the domestic shipping industry.

3. Selective Deregulation Captured by the Industry

One of the pieces of evidence for a neoliberalism explanation for the Sewol tragedy was the extension of the passenger ship's age limit from 25 to 30 years introduced under the conservative Lee Myung-bak government (2008–2012). Indeed, the Cheonghaejin would not have purchased the 18-year-old vessel retired in Japan in the first place in the absence of this deregulation. However, it is difficult to explain the overall trend in the regulatory regime for the domestic shipping industry with a neoliberal logic. Had the Korean government truly followed neoliberal ideas, the first deregulation should have been to remove the entry regulation that had protected monopolies in the ferry industry (Choi 2014; You and Park 2017).

Despite the rhetoric to promote neoliberal policies, the Korean government continued to protect monopolies in the ferry industry through an entry regulation. Out of the 99 domestic coastal ferry routes, as many as 85 routes were monopolized by a single shipping company at the time of the Sewol ferry disaster. Among them was the owner of the Sewol ferry enjoying a lucrative voyage license for the popular Incheon–Jeju route for 20 years (Shin 2014). The Marine Industry Act requires a business license for each ferry route, and in return for protecting the monopolies, the government imposes a cap on passenger fees (Choi 2014).

While the ferry industry complained the passenger fee was capped so low that they could not make a profit, the industry was apparently enjoying the state-sponsored cartel system. Considering the increasing competition with low-cost flights, it would have been difficult to charge substantially higher prices even without the price cap. In addition, the government was providing many tangible and intangible subsidies to the industry and individual firms like Cheonghaejin. For example, the Jeju Regional Marine Environment Management Corporation sent out official letters to Provincial Offices of Education in metropolitan areas to encourage high schools under their jurisdiction to organize ferry trips to Jeju Island as their annual student excursion package (Woo 2014: 105–112). This episode helps us to understand why so many victims of the Sewol were high-school students and how the public corporation maintained such a close relationship with private business as to conduct an effective marketing operation on behalf of Cheonghaejin.

While the government did not remove obsolete economic regulations such as the entry regulation and the price regulation, it was more accommodating to the industry's demand for deregulation in the area of safety regulations such as the passenger ship age limit. In particular, under the business-friendly Lee Myung-bak administration (2008–2012), it was easier for the industry to persuade the government to extend the maximum ferry life. During the policy-making process, the Anti-Corruption and Civil Rights Commission reported at a cabinet meeting that such deregulation would not risk safety hazards but would rather be economically effective for the passenger shipping business (Lawyers 2014: 69–70). An industry-commissioned research finding that most advanced countries did not have a regulation on passenger ships' age limit was conveniently cited, but the fact that these countries were applying the ISM code was simply ignored (You and Park 2017). When the maximum age of vessels was extended from 25 to 30 years in 2009, no supplementary measures to ensure shipping safety were introduced. An outcome of the industry-driven deregulation at the expense of public safety is well reflected in the surge of ferry accidents from 13.6 per year under the Roh Moo-hyun government (2003–2007) to 17 under the Lee Myung-bak government (2008–2012), a 25% increase (Lawyers 2014: 72–78). The sinking of the Sewol was part of this surge.

As such, the shipping industry was enjoying the better of two possible worlds. When deregulation and privatisation suited its interest, the shipping industry utilized the rhetoric of neoliberalism to justify these measures on the one hand while continuing to enjoy the state protection of monopolies on the other hand. The industry benefited from both the regulation (market entry) and the deregulation (ship age limits) that were directly shaped by its influence at the expense of public interest.

The Enduring Legacies of State Corporatism and Regulatory Capture

In this section, we trace the process by which regulatory capture has been institutionalized. Our investigation shows that the delegation of regulatory authorities to private entities was introduced in the context of state-corporatism of the authoritarian developmental state. Despite a series of liberalization and reform efforts, the Korean government has failed to bring under control the collusive practices such as parachute appointments that had been established under the state-corporatist arrangements.

1. Delegation of Safety Regulatory Authority to Private Entities

Although many pundits consider the vessel-operational managers system run by the KSA a form of privatization driven by a neoliberal policy, the system was first introduced a lot earlier than the influx of neoliberalism in Korea, after the outbreak of the Namyoung Ferry tragedy that took 338 lives in 1970. Because the accident was caused by overloading and overcrowding on board, the government decided afterwards to introduce vessel-operational managers charged with the responsibility to inspect the vessels' compliance with safe operational rules (Ju et al. 2009; KSA 2012). To avoid spending an extra budget on this new program, the government decided to delegate to the Korean Shipping Association the authority to hire vessel safety operational managers and bear their expenses. The Park Chung-hee government accordingly amended the Marine Transportation Industry Act in 1972.

The corporatist regulatory system could be implemented thanks to the existence of a peak industry association—the KSA—controlled by the government. The KSA had been established under the Korean Shipping Association Act of 1961 as part of President Park's reorganization of the state–society relationship according to the principles of state corporatism (KSA 2012: 147). The KSA has since been singled out as the only legitimate association representing the interest of the shipping industry, with more than 2000 companies and ship owners under its authority today. Over time, the KSA has complained about the financial burden imposed on member companies to pay for the vessel-operation managers system. When the ISM was being considered, however, the industry association successfully lobbied the government to maintain the vessel-operation managers system in order to thwart the introduction of the ISM code that would incur higher costs. The corporatist self-regulatory system in which the regulators (vessel operation managers) and the regulated (member shipping companies) were under the same KSA was inherently unworkable, even under the strong state, as attested by the continuous passenger ship accidents due to overloading and overcrowding, such as the sinking of the Dongyang Jumbo Ferry in 1984 and the Seohae Ferry in 1994.

While the delegation of the inspection authority to the Korean Register (KR), a not-for-profit ship classification society, is often held up as another piece of evidence of neoliberal privatization, the arrangement dates back to the 1960s, long before neoliberalism itself. The KR has held the authority for safety inspections and classification for certain ships since 1961 and its authority has been extended to passenger ships since 1982 under the Ship Safety Act. In fact, the delegation of the regulatory enforcement and inspection authorities was prevalent under Park Chung-hee and Chun Doo-hwan regime's state corporatism (Han 1989). The corporatist delegation did not mean the state's withdrawal from intervention in the economy because it was based on the state's tight control of the nominally private entities.

Furthermore, our research shows that there was no privatization of rescue operations, contrary to some commentators' perception that rescue operations were delegated to a private entity, the Maritime Rescue and Salvage Association of Korea (MARSA). The Coast Guard, rather than the MARSA, was still in charge, and it only arranged a lucrative contract for salvage operations between Cheonghaejin and Undine, a salvage company the CEO of which was vice-president of the MARSA.

The MARSA was created as a non-profit corporate according to the Rescue and Aid at Sea and River Act amended in 2012. The Coast Guard, as the main government agency to implement the Act, and relevant private groups had lobbied the government to legally recognize a non-profit organization to assist the Coast Guard with rescue and salvage operations (Koo 2014). Their rationale was that a rapid increase in the number of accidents at sea called for more participation from volunteer rescuers and non-profit rescue organizations, and an efficient system was needed to facilitate public–private cooperation. According to the Act, the primary purpose of the MARSA is "research, development, and training" on maritime rescue and salvage. The law allows the government to contract out some rescue and salvage works to the organization.

While the main rationale for the MARSA was to facilitate the participation of volunteers and non-profit rescue organizations, its leadership positions were occupied by a number of CEOs of marine companies and a few Coast Guard officials. It functioned as a channel for exchange of profitable information and favors (You and Park 2017). While the main benefits to the marine companies were lucrative information and even contracts (such as a salvage contract in the case of Sewol) through this collusive relationship, what were the main benefits to Coast Guard officials? The answer might be found in the opportunities for parachute appointments that we will discuss further in the next section.

2. The Enduring Legacies of State Corporatism

There has been a series of economic liberalizations since the 1980s, and this process has accelerated since the 1990s. Also, several attempts were made to reform collusive government–business relations including the practice of parachute appointments and the corporatist business associations. However, these attempts largely failed. Once collusive relations were established between the bureaucrats and the businesses under state corporatism, it was all but impossible to overcome the resistance from vested interests.

The practice of parachute appointments originated from the state-corporatist arrangements of the Park Ching-hee period. The state tightly controlled the leadership selection of business associations. Often the state arranged to give leadership posts to retired government officials, and the associations accepted these parachute appointments as a channel of communication and influence to the government (Park 1987). Moreover, the practice spread to post-retirement appointments in private businesses.

As widespread practice of parachute appointments strengthened collusive ties between the government and business and facilitated corruption, public criticism rose. In response, the Chun Doo-hwan government enacted Public Officials Ethics Act in 1981, which prevented senior officials from taking a job for two years after retirement in private companies in the sectors to which their previous work was related. The law did not prohibit retired officials from taking positions in non-profit organizations such as industry associations and various interest groups. And this loophole was widely utilized, rendering the employment ban largely ineffective.

As part of the Kim Dae-jung administration's anti-corruption reforms in the aftermath of the financial crisis, the National Assembly amended the Public Ethics Act in 2001 to ban retired senior officials from being employed in not-for-profit organizations established by private companies for their common interests, as well as profit-seeking companies. Once again, however, the legal amendment fell short by exempting organizations that conducted works contracted out by central or local governments. This loophole was created by the presidential enforcement decree, and senior bureaucrats seem to have influenced the drafting of the decree. Thus, retired bureaucrats continued to enjoy lucrative positions in such organizations, given that a number of government works had been delegated or contracted out to semi-public and private entities in not-for-profit organizational forms including business associations.

In addition, this exception clause further incentivized bureaucrats to create more such organizations. It was in this context that the MARSA was created in 2012 as a legally recognized non-profit organization eligible for government-delegated rescue and salvage-related operations. The primary benefit of the MARSA to Coast Guard officers was to provide them with additional job opportunities after retirement (You and Park 2017).

Another area that shows the enduring legacy of corporatist practices of the authoritarian period is industry associations. The Regulatory Reform Committee of the Kim Dae-jung government prioritized the task of transforming the singular, compulsory, and uncompetitive characteristics of the business associations into more plural, autonomous, and competitive arrangements. SaKong and Kang (2001) show, however, the results of this reform initiative were minimal because of strong resistance from both bureaucrats and business associations.

Comparison with the Fukushima Nuclear Accident

As we noted earlier, the Fukushima Nuclear Accident Independent Investigation Commission's official report concludes that the man-made accident was the "result of collusion between the government, the regulators and Tokyo Electric Power Company (TEPCO), the operator of the Fukushima Daiichi Nuclear Power Plant" (National Diet of Japan 2012). It declares that the relationship between the regulators and operators was "a typical example of regulatory capture." Although both Japan's Nuclear and Industrial Safety Agency (NISA) and the operators were aware of the risk of core damage from a tsunami for a long time, "no regulations were created, nor did TEPCO take any protective steps against such an occurrence," according to the report. The lack of independence, transparency, expertise, and professionalism of the regulatory body made it prone to capture by the nuclear industry. In particular, the report notes that Japan's nuclear regulatory capture stems from its regulatory structure. NISA is a division of the Ministry of Economy, Trade and Industry (METI), which is responsible for promoting the nuclear power industry. There is an inherent "promoter-regulator conflict," which makes NISA prone to industry capture.

We find an analogy between the Sewol and the Fukushima accidents regarding the fundamental cause of regulatory capture. As in Japan's nuclear safety regulation, the regulatory structure itself created regulatory capture in the marine safety regulation in Korea. In fact, the Korean case was even worse than the Japanese case because its regulatory enforcement authority was granted to the shipping industry itself. As Japan's Federation of Electric Power Companies (FEPC) "manipulated the cozy relationship with the regulators to take the teeth out of rules and regulations," the Korean Shipping Association did the same (National Diet of Japan 2012). Regulatory capture led to inadequate regulations in both cases, failing to meet international standards. In Japan, the laws and regulations on nuclear safety were based on the assumption that the scope and magnitude of possible natural disasters would not exceed precedent, without taking into account the prospect of unprecedented events. Japanese laws and regulations failed to fully incorporate the defense-in-depth concept, which was adopted in many advanced countries such as the United States and France. Likewise, domestic passenger shipping was exempt from the application of International Safety Management Code in Korea.

Another striking similarity between the two cases is the ubiquitous revolving-door practice that led to collusion and capture. Wang and Chen (2012) argue that no industry is perhaps as rife with revolving-door practices, such as Amakudari (descent from heaven; equivalent to parachute appointments in Korea) and Amaagari (ascent to heaven; moving from the private sector to government agencies), as the nuclear power sector in Japan. They note that four former senior officials from the nuclear regulatory agencies served as vice presidents of TEPCO from 1959 to 2010. Out of 95 people at the three main nuclear regulatory bodies (NISA, Atomic Energy Commission, and Nuclear Safety Commission of Japan) at the time of the Fukushima accident, 26 had been affiliated with the nuclear industry or groups that promoted nuclear power.

In the aftermath of Japan's Fukushima nuclear disaster, a number of studies discuss the issue of regulatory capture and the iron triangle of the pro-nuclear bureaucrats, politicians, and power industry (Howe and Oh 2013; Synolakis and Kânoglu 2015; Wang and Chen 2012; Yokoyama 2013), but these studies focus on the consequences of regulatory capture, neglecting its causes. In South Korea, likewise, some studies discuss the collusive relations between the ferry industry and those in charge of safety regulation for passenger shipping, but they fall short of exploring the causes of regulatory capture (Hong 2014; Roh 2014).

Why did the Japanese and Korean regulators maintain cozy relations with the regulated industry? Was their regulatory capture inevitable or preventable? What institutional mechanisms and policy measures can help prevent regulatory capture? These are critically important questions to explore in order to prevent man-made catastrophic accidents in the future, not just in Korea or Japan but also in the rest of the world. Although it is beyond the scope of this study to comprehensively explore these questions, it is worth noting that corporatist legacies have contributed to making regulatory agencies prone to capture in both countries.

The likelihood and degree of regulatory capture will be substantially determined by the incentives and abilities of the regulated industry (or special interests) to capture the regulators, which will be affected by the market structure of the industry as well as government–business relations. Industries with perfectly competitive markets with perfect information will have few incentives to capture the regulators. A corporatist structure of interest mediation in government–business relations will be more prone to capture than a pluralist structure. Industrial policy will also be a significant factor. The prevalence of revolving door practices is also likely to be correlated with the likelihood and degree of capture. Both the developmental states of Japan and Korea, or "Japan, Inc." and "Korea, Inc.," adopted corporatism in the organization of government–business relations. Japan replaced the pre-war state corporatism with societal corporatism without labor in the postwar period (Pempel and Tsunekawa 1979), and Korea emulated Japanese pre-war system to develop its state corporatism. Their corporatist structures have left enduring legacies that still define government–business relations to a considerable extent.

Conclusion

We have thus far shown that the serious failures in the safety regulation manifested in the case of the Sewol ferry were largely caused by regulatory capture. While neoliberal deregulations such as the extension of passenger ships' maximum life also contributed to the Sewol accident, a more fundamental problem is the regulatory capture that has been institutionalized as part of the enduring legacy of Korea's state corporatism. This should call for the deregulation of the sectors that are still under bureaucratic control and protection, effectively maintaining a state-sponsored cartel system.

Regulatory capture has distorted regulatory policy-making and weakened regulatory enforcement. The continuing regulation of market entry that protects the state-sponsored cartel system and the deregulation of safety measures indicate that the priorities for regulatory policies were not guided by economic logic but driven by the vested interests of the industry and bureaucrats, which prevent a reform of the old system. In retrospect, there were missed opportunities, such as the failed attempts to replace the ineffective vessel-operational managers system with the ISM code and to make the vessel-operational managers system independent from the KSA.

After the Sewol incident, a number of reforms have been implemented, including the dissolution of the Coast Guard, the creation of the Ministry of Public Safety and Security, and the transfer of the affiliation of vessel-operational managers from the KSA to the Korea Ship Safety Corporation as the failed legislative bill in 2011 had proposed. However, it is questionable whether these reforms will be sufficient to break up the existing regulatory capture and improve marine safety, for they fall short of dismantling the cartel system in the coastal ferry industry. There have been no systemic reforms to empower the consumers and various stakeholders in the realm of safety regulations either (You and Park 2017).

Considering our study of the Sewol ferry accident and a number of studies on the Fukushima nuclear accident identify regulatory capture as an important cause of catastrophic man-made disasters, there is a need for a more comprehensive study of the causes and remedies of the phenomenon. While our study shows that the legacies of state corporatism significantly contributed to institutionalizing regulatory capture in Korea, further studies will be needed to better account for what elements of corporatism defy liberalization or democratization and how they are institutionalized into regulatory capture. We will then be able to better control and prevent regulatory capture—and help create a safer society.

References

Bennett, A., & Checkel, J. T. (Eds.). (2015). *Process tracing: From metaphor to analytic tool.* Cambridge, New York: Cambridge University Press.

Board of Audit and Inspection (BAI). (2014). Interim report on Sewol ferry accident (세월호 침몰사고 대응실태 중간발표 내용). *Journal of Korean Maritime (해양한국), 2014*(8), 109–115.

Carpenter, D. (2014). Detecting and measuring capture. In D. Carpenter & D. A. Moss (Eds.), *Preventing regulatory capture: Special interest influence and how to limit it.* New York, NY: Cambridge University Press.

Carpenter, D., & Moss, D. A. (Eds.). (2014). *Preventing regulatory capture: Special interest influence and how to limit it.* New York, NY: Cambridge University Press.

Cawson, A. (1986). *Corporatism and political theory.* Oxford, UK: Basil Blackwell Ltd.

Choi, B. S. (최병선). (2014). Regulatory reform after the sewol tragedy (세월호 참사 이후 규제개혁의 진로). *Korea Public Administration Forum* (한국행정포럼), *145*, 10–13.

Choi, J. J. (1989). *Labor and the authoritarian state: Labor unions in South Korean manufacturing industries, 1961–1980.* Seoul: Korea University Press.

Dal Bó, E. (2006). Regulatory capture: A review. *Oxford Review of Economic Policy, 22*(2), 203–225.

Green, P. (2005). Disaster by design: Corruption, construction, and catastrophe. *British Journal of Criminology, 45*(4), 528–546.

Haggard, S. (2000). *The political economy of the Asian financial crisis.* Washington, DC: Institute for International Economics.
Han, Y. S. (한영석). (1989). A study on the delegation of administrative authority (행정권한의 위임에 관한 고찰). *Choong-Won Review of Humanities* (중원인문논총), *8*, 87–104.
Hong, S. T. (홍성태). (2014). *Assessment of risk society: Going beyond accident society to a safe society (위험사회를 진단한다: 사고사회를 넘어 안전사회로).* Seoul: Aropa (아로파).
Hong, S., & Lim, J. (2016). Capture and the bureaucratic mafia: Does the revolving door Erode bureaucratic integrity? *Public Choice, 16*(1), 69–86.
Horiuchi, A., & Shimizu, K. (2001). Did amakudari undermine the effectiveness of regulator monitoring in Japan? *Journal of Banking & Finance, 25*(3), 573–596.
Howe, B. M., & Oh, J. S. (2013). The Fukushima nuclear disaster and the challenges of Japanese democratic governance. *Korea Observer, 44*(3), 495–516.
Jang, H. S. (장하성). (2014). 한국자본주의: 경제민주화를 넘어 정의로운 경제로 (*Korean capitalism: Toward a just economy beyond economic democratization*) (서울: 헤이북스). Seoul: Heybooks.
Ji, J. H. (지주형). (2013). 1997 년경제위기 이후 한국의 신자유주의화: 위기담론과 위기관리의 문화정치학 (Neoliberalization of South Korea after the 1997 economic crisis: A cultural political economy of crisis discourse and management). 한국정치학회보*(Korean Political Science Review)*, *47*(3), 33–58.
Ji, J. H. (지주형). (2014). Political science approach to Sewol disaster: Idealism of neoliberalism and its reality (세월호 참사의 정치사회학: 신자유주의의 환상과 현실). *Economics and Society (*경제와 사회*)*, *104*, 14–55.
Ju, J. K., Jo, I. H., Choi, S. Y., & Lee, E. B. (주종광, 조인현, 최석윤, 이은방). (2009). *A study on the developing solutions of management system on coastal passenger ships* (내항여객선 운항관리제도 발전방안 고찰). Association for the Study of Marine Environment and Safety, Fall 2009 Conference (해양환경안전학회 2009년도 추계학술발표회).
Kalinowski, T. (2008). Korea's recovery since the 1997/98 financial crisis: The last stage of the developmental state. *New Political Economy, 13*(4), 447–462.
Kim, E. Y. (김의영). (2005). A study of corporate interest representation system in Korea: Focusing on the relations between the state and business associations (한국의 기업이익대표체제에 대한 소고: 국가와 사업자단체의 관계를 중심으로). *Korea-Japan Joint Research Series (*한일공동연구총서*)*, *9*, 156–182.
Kim, S. J. (김상조). (2012). 종횡무진 한국경제(*A thorough analysis of the Korean economy*) (서울: 오마이북). Seoul: Oh My Book.

Kim, Y. R. (김영래). (1987). *Interest groups in Korea: From the state-corporatist perspective (한국의 이익집단: 국가조합주의적 시각을 중심으로)*. Seoul: Daewangsa (대왕사).
Kim, Y. T. (1998). The State, Capital and Labor: Korean corporatism in comparative perspective. *The Journal of Asiatic Studies (아세아연구), 41*(2), 187–215.
Kong, T. Y. (2004). Neo-liberalization and incorporation in advanced newly industrialized countries: A view from South Korea. *Political Studies, 52*, 19–42.
Koo, Y. H. (구용회). (2014). Coast guard-MARSA: Rort connection (해경-해양구조협회 '추악한커넥션' 의혹). *Nocut News (노컷뉴스)* 1 May. Retrieved November 23, 2014, from http://www.nocutnews.co.kr/news/4016537. Accessed on November 23, 2014.
Korean Shipping Association (KSA) (한국해운조합). (2012). *Fifty-year history of the Korean shipping Association* (한국해운조합50년사). Seoul: Korean Shipping Association.
Lawyers for Democratic Society (민주사회를위한변호사모임). (2014). *416 Sewol ferry transcripts of the lawyers of democratic society (세월호민변의기록)*. Seoul: Path of Thoughts Publications (생각의길출판사).
Lee, J. S. (이진순). (2003). *Korean economy: Crisis and reform* (한국경제: 위기와개혁). Seoul: Book 21.
Lee, Y. (2015). "The Sewol disaster: Intended consequences of neoliberal deregulation" Draft.
Lewis, J. (2011). Corruption: The hidden perpetrator of under-development and vulnerability to natural hazards and disasters. *Journal of Disaster Risk Studies, 3*(2), 464–475.
Lim, H. (2010). The transformation of the developmental state and economic reform. *Journal of Contemporary Asia, 40*(2), 188–210.
Lim, H. C., & Jang, J. H. (2006). Between neoliberalism and democracy: The transformation of the developmental state in South Korea. *Development and Society, 35*(1), 1–28.
Makkai, T., & Braithwaite, J. (1992). In and out of the revolving door: Making sense of regulatory capture. *Journal of Public Policy, 12*(1), 61–78.
McNamara, D. L. (1999). Korean capitalism. In D. L. McNamara (Ed.), *Corporatism and Korean capitalism*. New York: Routledge.
Moon, C., & Mo, J. (2000). *Economic crisis and structural reforms in South Korea: Assessments and implications*. Washington, DC: Economic Strategy Institute.
National Diet of Japan. (2012). *The official report of the Fukushima nuclear accident independent investigation commission*.
Oh, C. R. (오창룡). (2014). Sewol disaster and irresponsible governance: Incompetence of state power and neoliberalism (세월호 참사와 책임회피 정

치: 신자유주의 국가권력의 무능전략). *Progressive Review (진보평론)*, *61*, 37–52.
Onis, Z. (1991). The logic of the developmental state. *Comparative Politics, 24*(1), 109–126.
Park, M. K. (1987). Interest representation in South Korea: The limits of corporatist control. *Asian Survey, 27*(8), 903–917.
Park, Y. S. (2011). Revisiting the South Korean developmental state after the 1997 financial crisis. *Australian Journal of International Affairs, 65*(5), 590–606.
Pempel, T. J., & Tsunekawa, K. (1979). Corporatism without labor? The Japanese anomaly. In P. C. Schmitter & G. Lehmbruch (Eds.), *Trends toward corporatist intermediation*. London: Sage Publications Inc.
Pirie, I. (2012). The new Korean political economy: Beyond the models of capitalism debate. *The Pacific Review, 25*(3), 365–386.
Roh (노진철), J. C. (2014). Structural problems of Sewol disaster and limitations of rescue management system (세월호 참사의 사회구조적 원인과 재난대응 체계의 한계). *Journal of Korean Maritime (해양한국), 6*, 138–150.
SaKong, Y. H, & Kang, H. W. (사공영호, 강휘원). (2001). The corporatist interest representation system of business associations and its costs (사업자 단체의 조합주의적 이익대표체제와 그 비용). *Korean Public Administration Review (한국행정연구), 10*(1), 157–81.
Salant, D. (1995). Behind the revolving door: A new view of public utility regulation. *Rand Journal of Economics, 26*(3), 362–377.
Schmitter, P. C. (1979). Still the century of corporatism? In P. C. Schmitter & G. Lehmbruch (Eds.), *Trends toward corporatist intermediation*. London: Sage Publications Inc.
Seo, Y. J. (서영재). (2014). "Problems of the vessel safe-operational management system for coastal ferry business and possible remedies (내항여객운송 사업 안전운항관리제도의 문제점 및 개선방안)" National assembly research service (국회입법조사처), *Issues and Controversies* (이슈와 논점), No. 897 (August, 12, 2014).
Shin, I. H. (신익환). (2014). The Cheonghaejin, full of suspicious dealings (까면 깔수록, 청해진해운 의혹투성이), *NewsTomato*. April 24.
Soliman, H., & Cable, S. (2011). Sinking under the weight of corruption: Neoliberal reform, political accountability and justice. *Current Sociology, 59*(6), 735–753.
Stigler, G. (1971). The theory of economic regulation. *Bell Journal of Economics and Management Science, 2*, 3–21.
Stubbs, R. (2009). What ever happened to the east asian developmental state? The unfolding debate. *Pacific Review, 22*(1), 1–22.
Suh, J. J. (2014). The failure of the South Korean national security state: The Sewol tragedy in the age of neoliberalism. *The Asia-Pacific Journal, 12*, 40.

Synolakis, C., & Kânoğlu, U. (2015). The Fukushima accident was preventable. *Philosophical Transactions of the Royal Society A, 373,* 20140379.

Um, K., Lim, H. C., & Hwang, S. M. (2014). South Korea's developmental state at a crossroads: Disintegration or re-emergence. *Korea Observer, 45*(20), 211–253.

Unger, J., & Chan, A. (1995). China, corporatism, and the East Asian model. *The Australian Journal of Chinese Affairs, 33,* (Jan.), 29–53.

Wade, R. (1990). *Governing the market: Economic theory and the role of government in East Asian industrialization.* Princeton, NJ: Princeton University Press.

Wang, Q., & Chen, X. (2012). Regulatory failures for nuclear safety-the bad example of Japan: Implication for the rest of world. *Renewable and Sustainable Energy Reviews, 16*(8), 2610–2617.

Woo, S. H. (우석훈). (2014). *Trapped inside the ferry (내릴수없는배).* Seoul: Woongjin House (웅진지식하우스).

Yeung, H. (2014). Governing the market in a globalizing era: Developmental states, global production networks and inter-firm dynamics in East Asia. *Review of International Political Economy, 21*(1), 70–101.

Yokoyama, J. (2013). Fukushima disaster and reform. *Environmental Policy and Law, 43*(4/5), 226–233.

You, J. S. (2015). *Democracy, inequality and corruption: Korea, Taiwan and the Philippines compared.* Cambridge, U.K.: Cambridge University Press.

You, J. S., & Park, Y. M. (2017). The legacies of state corporatism in Korea: Regulatory capture in the Sewol ferry tragedy. *Journal of East Asian Studies, 17*(1), 95–118.

You, J. S. (Forthcoming). Demystifying the Park Chung-Hee Myth: The critical role of land reform in the evolution of Korea's developmental state. *Journal of Contemporary Asia.*

Young, B. (2011). "Neoliberalism". In B. Badie, D. Berg-Schlosser, & L. Morlino (Eds), *International encyclopaedia of political science.* Thousand Oaks: SAGE Publications, Inc.

Authors' Biography

Jong-sung You (유종성) is Senior Lecturer at the Department of Political and Social Change, Australian National University. In addition to the book, *Democracy, Inequality and Corruption: Korea, Taiwan and the Philippines Compared* (Cambridge University Press 2015) (동아시아 부패의 기원: 문제는 불평등이다. 한국 타이완 필리핀 비교연구. 동아시아출판사 2016) his publications appeared in *American Sociological Review, Political Psychology, Journal of Politics, Journal of East Asian Studies, Journal of Contemporary Asia, Asian Perspective, Trends and Prospects, and Korean Journal of International Studies.* He obtained a Ph.D. in Public Policy and MPA from Harvard University. He was General Secretary of Citizens' Coalition for Justice.

Youn Min Park is a visiting fellow in the Department of Political and Social Change at the Australian National University. With a focus on the concept of information/knowledge in the digitally networking society, Dr. Park extends her research interest to political mobilization in Korean society, everyday life politics, and freedom of speech. Her research has appeared in peer-reviewed journals, including *Development and Society, International Review of Public Administration* and *Journal of International and Area Studies*.

CHAPTER 6

"Stay Still": Sewol, a Tale of Fatal Censorship, Fatal Paternalism

K. S. Park

Accidents can happen anywhere, anytime. In democracies, people are allowed to make a wider range of choices and therefore assume greater risks. An accident like Sewol can happen in a welfare state,[1] as well as a republic newly freed from a socialist state.[2] What separates an ordinary accident from a mass disaster of a socially catastrophic nature is whether proper rescue efforts are implemented once an accident happens, and if not, whether there are good reasons for the failures. We can make this enquiry further upstream in causation. Corruption and the resulting anomalies, and mad drives to profit are the causes of these accidents. However, greed, and also government–business intermingling to a lesser extent, is part and parcel of the capitalistic market economy. What separates a mere risk from its materialization is whether people are allowed to raise the proverbial "red flag" when they see safety-threatening corruption or profiteering.

[1] http://news.bbc.co.uk/2/hi/europe/539764.stm.
[2] https://en.wikipedia.org/wiki/MS_Estonia.

K.S. Park (✉)
Korea University, Seoul, South Korea

J.-J. Suh and M. Kim (eds.), *Challenges of Modernization and Governance in South Korea*, DOI 10.1007/978-981-10-4023-8_6

The catastrophic failure to respond to a *known crisis* (or a known risk, as to be discussed later) shown in Sewol highlights the importance of these questions. The Sewol accident was a social catastrophe. A ship submerged slowly over a span of 2 h right within a stone's throw of the Coast Guard ship in a relatively warm and calm ocean near the coast, and the ship then sank further down with the 300 or so then survivors locked inside the ship, again slowly over a span of three days, while its descent was being televised real-time to the whole country. The Sewol was no ordinary mass disaster that could be attributed to a lapse of judgment or two: it was a tragic march toward certain death over a sustained period of time, during which several clear opportunities for restoring normalcy went ignored one by one. The end result could have been well expected from a series of pre-meditated non-actions. One of the most popular yet disturbing slogans surrounding Sewol was: "Capital sank the Ship, and the State caused the Massacre."[3] One can say that failure to respond to the known crisis had a more deterministic impact than the negligent lapses of judgment that initially caused the crisis. No matter how distributively efficient or morally competent the government is, a Sewol can happen again. How the authorities, the government, respond to it shall be the unique lessons learned from the Sewol tragedy.

This essay reports in Part I (Fatal Censorship) on the key events before and surrounding the Sewol ferry sinking and finds that, on both of the above questions, the answer has been a resounding "No", punctuated by the fatal warning of "Stay Still", which in the Korean original means both "stay still" and "stay silent".

Of course, the most mindboggling question is why the captain and crew gave only a "stay still" order and never followed up with an "abandon ship" order even as the ship tilted to a fatal angle while the passengers consisting of mostly high-school students obediently waited; and likewise why the Coast Guards failed to intervene during a series of opportune moments. This essay reexamines the causes of other disasters of similar social significance in the past and attempts at conjectures on the relationship between paternalism and mass disasters in Part II (Fatal Paternalism).

---

[3] http://www.ohmynews.com/NWS_Web/View/at_pg.aspx?CNTN_CD=A0002101095; http://wspaper.org/7_print_popup.php?articleNos=15622%2C15623%2C15625&title=%EC%84%B8%EC%9B%94%ED%98%B8+%EC%B0%B8%EC%82%AC+1%EB%85%84%EC%9D%84+%EB%8F%8C%EC%95%84%EB%B3%B4%EB%A9%B4%EC%84%9C.

Fatal Censorship

Before the Accident

In January 2014, a former employee of Chonghaejin Marine Transport, the operator of Sewol Ferry, terminated with unpaid wages, logged on www.epeople.go.kr to report the repeated overloading on the ferry to the President's Office, as well as his back-wage claims.[4] Unfortunately, the Prime Minister's Office, which in fact handles complaints, resolved only his wage claims but did not take any action on the overloading. The newly opened site received a large number of requests for help and the Prime Minister's Office focused on what seemed to be the more personally dire problems for the complainant.

One cannot help but wonder why the former employee could not do what many former employees do: publicly demand the back wages and also criticize the employer about the overloading, which would have attracted sufficient public attention to resolve his wage claims quickly. "Normalization of deviance" is indeed rampant, as the country's new legal and regulatory systems are still experimenting with various standards later found to be impractical and largely ignored by both the regulators and their subjects.[5] However, when it comes to unpaid wages, public resentment is universal and had a good chance of bringing the accompanying problem of overloading into public scrutiny, thereby averting the disaster.

However, in Korea, everyone reporting truthfully on another's crime risks becoming a criminal himself or herself. First of all, the crime of insult, criminal defamation and "truth defamation" laws are still being vigorously enforced in Korea, despite the warnings of international human rights bodies, including General Comment No. 34 of the Human Rights Committee, which condemned incarceration as punishment for defamation and penalization of truthful statements.[6] Criminal prosecution continues to be a prevailing remedy for defamation or insult. Over a

[4] http://www.hani.co.kr/arti/society/society_general/634965.html%20.

[5] See Vaughan, D. (1996). The Challenger launch decision: risky technology, culture, and deviance at NASA. Chicago: University of Chicago Press (Vaughan 1996).

[6] Para. 47

55 months period between January 1, 2005 and July 2009 in Korea,[7] 136 people were incarcerated for defamation or insult, while in comparison only 146 people have been incarcerated for defamation in a 20-month period between January 1, 2005 and August 2007 in all other countries combined.[8] On average, Korea accounted for about 30% of the worlds' defamation incarcerations in that 20-month period.

The trend continues to date and with greater intensity. In 2013, 11,579 people were indicted for defamation or insult (2,162 for defamation and 9,417 for insult, excluding 1,233 indicted for online defamation),[9] out of which 111 were incarcerated while the remaining defendants were fined.[10] This is a twofold increase from 2010: a total of 7,058 people (2,198 for defamation and 4,860 for insult)[11] were indicted, out of which 43 incarcerations resulted. [12]

As Special Rapporteur Frank La Rue pointed out in his report on Korea, many of these criminal prosecutions are the very cases where private persons are subjected to criminal prosecution for defamation in defense of public officials' reputation.[13] We are certain of the political nature of these prosecutions because most cases result in withdrawal, dismissal, or not-guilty judgments, leaving only the indelible chilling effects on the populace.[14]

Secondly, such abuse is facilitated by the fact that criminal prosecution applies also to truthful statements (or statements not proven to be

---

[7] MP LEE Chun-Seok's Press Release, October 19, 2009 http://media.daum.net/tvnews/view.html?cateid=100000&newsid=20101006161113668&p=newsis.

[8] http://www.article19.org/advocacy/defamationmap/overview.html [no longer available; last accessed on May 30, 2009].

[9] Prosecutors' Office Year Book of 2014, Chapter 6, Pages 926, 966 http://www.spo.go.kr/spo/info/issue/spo_history02.jsp?mode=view&board_no=64&article_no=590945.

[10] Courts' Year Book of 2014, Section on Crimes, Chapter 5, Page 89 http://www.scourt.go.kr/portal/justicesta/JusticestaListAction.work?gubun=10.

[11] Prosecutors' Office Year Book of 2011, Chapter 6, Pages 1048, 1049. http://www.spo.go.kr/spo/info/issue/spo_history02.jsp?mode=view&article_no=518576&pager.offset=0&board_no=64&stype=&sort_year=2011.

[12] Courts' Year Book of 2011, Chapter 5, Pages 874, 875.

[13] Report of the Special Rapporteur on the promotion and protection of the right to freedom of opinion and expression, Frank La Rue, on his mission to the Republic of Korea (May 6–17, 2010), A/HRC/17/27/Add.2, paras. 25, 89 http://daccess-dds-ny.un.org/doc/UNDOC/GEN/G11/121/34/PDF/G1112134.pdf?OpenElement.

[14] 참여연대, <국민입막음 소송 보고서> (2015).

false)—even in the absence of privacy concerns, in contrast to the Special Rapporteur's[15] and UN Human Rights Committee's[16] specific mandates to exempt such statements. The defendant can only escape liability by proving that the statements were made solely for public interest, a burden of proof not so easy to sustain. For instance, the Supreme Court refused the public-interest defense of a worker making a truthful statement about his employer's non-payment of wages, as the court found that the worker's statement was also intended to harm the employer's reputation, i.e., the public interest was not the sole motive.[17] The same reasoning was applied to a drug wholesaler, who truthfully complained about the pharmaceutical companies' unfair trading practices.[18] The practical effect of this law has been that an individual who has discovered revealing truths about corruption in government or other powerful entities could not freely share them with others for fear that they may not be able to sustain the burden of proving that public interest was the speaker's sole motive.

Of course, some countries like Norway, the Netherlands, Denmark, Finland, and Switzerland[19] do retain truth defamation laws which require public interest as an element of defense but they apply it to disclosure of private facts, not to protect a malfeasor from loss of reputation. For instance, a law can easily be approved punishing disclosure of physical injury resulting from an accident[20] but not because a person's reputation is harmed but because their privacy is infringed upon people seeing their body parts. In Korea, the truth defamation law is actually used by a malfeasor to prevent people from talking about his or her malfeasance. For instance, a member of an elders association was found guilty of truth

[15] Special Rapporteur Frank La Rue's Korea Report, Para. 27 "The Special Rapporteur reiterates that for a statement to be considered defamatory, it must be false, must injure another person's reputation, and made with malicious intent to cause injury to another individual's reputation."

[16] General Comment 34, para. 47, "All ... penal defamation laws ... should include such defences as the defence of truth."

[17] Supreme Court 2004.10.15 Judgment 2004Do3912.

[18] Supreme Court 2004.5.28 Judgment 2004Do1497.

[19] Council of Europe, "Legal Provisions Concerning Defamation and Insult in Europe" http://www.coe.int/t/dghl/standardsetting/media/doc/dh-mm(2003)006rev_EN.asp.

[20] William Roos, Case Comment, NETHERLANDS: COPYRIGHT: RIGHT TO PRIVACY AND PORTRAIT RIGHT, Ent. L.R. 1998, 9(8), N146–147 (Roos 1998).

defamation when he alerted other members about violent behavior that the association's officer exhibited toward other members with no intention of keeping it private.[21]

Recently, a prominent poet was found guilty of "candidate slandering" for alleging that the then presidential candidate Park Gun-Hye had custody of calligraphies of Ahn Joong-Geun, an independence fighter who assassinated Ito Hirobumi, the Japanese prince that spearheaded the annexation of Chosun[22] where his allegations were not proven false. Although the judgment was fortunately later reversed for the reason that the poet had public interest in mind in making the claims,[23] this case demonstrated the legal risk that one has to face in revealing inconvenient truths about a powerful person.

During the Accident

The United Nations International Strategy for Disaster Reduction (UNISDR)'s Guide for Journalists Covering Disaster Risk Reduction starts out by saying, "The media can influence political decisions, change public attitudes and, of course, save lives".[24] Also, "a news organization plays four key roles during a crisis. It's a vital information resource, telling what is happening where, who is affected, how things are changing, and why. It's a communication lifeline, saving lives by relaying critical information to and from affected parties. It's an early warning beacon, transmitting timely, reliable information that prevents harm…"[25] The phrases above apply most poignantly to Sewol.

There was a curious interview broadcast on television a few days after the tragedy[26]: The father of a 16-year-old Sewol victim tearfully complained, "If the media gave a little more truthful and little more critical reports from the day the ship sank, I believe that more children would

---

[21] Supreme Court Judgment, March 28, 2013, 2012Do11914.

[22] https://www.koreatimes.co.kr/www/news/nation/2013/11/116_145842.html.

[23] http://www.yonhapnews.co.kr/politics/2014/03/25/0505000000AKR20140325076451055.HTML.

[24] Brigitte Leoni, Tim Radford, Mark Schulman, "Disaster Through a Different Lens: a guide for journalists covering disaster risk reduction" United Nations International Strategy for Disaster Reduction (UNISDR) (2011).

[25] Deborah Potter, Sherry Ricchiardi, "Disaster and Crisis Coverage", International Center for Journalists, p.8 (2002) http://www.icfj.org/sites/default/files/Disaster_Crisis.pdf.

[26] http://news.jtbc.joins.com/html/146/NB10471146.html?cloc=jtbc%7Cnews%7Cindex_photo_news.

have come back alive, and that belief has never been shaken. [...] the broadcasting closed their eyes during the most important two or three days. I believe this is the most important problem." What he meant was the tendency of the Korean media to deliver announcements by the rescue authorities without double-checking and their failure to examine their rescue efforts critically. The rank-and-file reporters of Munhwa Broadcasting Company issued a public apology "for not observing the basic principles of reporting... 'Repeating after' government's untrustworthy reports, we followed the habit of blurting out such phrases as '700 rescue personnel', '239 rescue boats', 'largest rescue operation', incongruent to the actual search situation [on the day of the accident] ... adding to the confusion in the urgent rescue situation"[27] while the first attempt to enter the sinking ship was made as late as 4 pm that day and the second one only the morning of the following day.

Actually, the media failure began as early as within a few hours of the sinking that had begun around 9 am.[28] Kim Hong-Kyung, a hero who saved 30 or so students by pulling them up the tilted floors of the lower deck using first makeshift ropes patched from the curtains and later a fire hose, before being rescued himself around 11 am, was interviewed by MBC and KBS around 4 pm while rescue efforts were still going on. What he said during the interview was unbelievable[29]: "While I was pulling up the kids using the curtains, the rescue team has not arrived. After they arrived, I pulled up about 15 more kids for about 30 min... The rescue team climbed on the outer fence of the ship but did not come down into the ship. I was curious to know why they were not coming in. They were just watching me from above pulling up the kids and repeatedly disappeared and reappeared... I was so curious that I even did selfies to show the Coast Guards just watching."

What was more unbelievable was that these comments were cut from the interview clips broadcast later by MBC and KBS.[30] These media

[27] A Statement of 121 Reporters of MBC, May 12, 2014; MBC Reporters' Association May 13, 2014.

[28] 여객선세월호 침몰사고 원인분석 결과보고서, 합동수사본부 전문가 자문단.

[29] http://m.media.daum.net/m/media/society/newsview/20140503102007922.

[30] "더 구할 수 있었는데..." 학생 20여명 살린 용감한 승객들(17일, 고아름 기자) KBS 9시 뉴스 http://news.kbs.co.kr/news/NewsView.do?SEARCH_NEWS_CODE=2847780&; MBC 뉴스데스크 침몰 순간에도 학생 먼저 구조(17일, 박진준 기자) http://imnews.imbc.com/replay/2014/nwdesk/article/3449691_13490.html.

organizations faithfully reported all the announcements of the rescue authorities but did not report on the crucial eye witness whose words could have begun the questioning about, and later possible revamping of, the rescue efforts on the crucial first day. Kim himself wondered, "The reporters just took all the pictures and videos but my interview on the Coast Guards' failure to assist was removed from the broadcast."

Now, what was more unbelievable was that a formal complaint was filed with the country's broadcasting content regulation authority, Korea Communication Standards Commission (KCSC), but KCSC decided that there was no problem with the reporting.[31]

This last incredible turn of events requires us to look at broadcasting regulation in Korea. KCSC is the administrative censorship body conducting the so-called "fairness review" of broadcasting content in Korea. Like the now defunct fairness doctrine in the USA, it requires that broadcasters "strive to provide an equal opportunity to other groups having different opinions, and also endeavor to maintain a balance in organizing the broadcast programs with respect to each political stakeholder."[32] Due to vulnerability to political abuses by successive regimes taking control of censorship bodies, and also due to the increasing importance of alternative media such as cable, fairness review by administrative bodies has disappeared in the USA, Japan, Germany, and the UK. Fairness review remains only in Korea, France, and the UK's non-BBC channels.[33]

In Korea, fairness review is alive and well and it has caused many controversies when KCSC disciplined broadcasters for not being fair to the legislation, personnel decisions, and foreign trade policies initiated by the government. Fairness review, as long as used to discipline broadcasting content for not being fair to the government, which in turn controls the reviewing body KCSC, eviscerates the fairness of the fairness review. In order to avoid a conflict of interest, the reviewing body should not discipline broadcasting content for being too critical of government-sponsored projects, laws, and ideas.

---

[31] 2014년 제27차 방송심의소위원회 회의록, 2014.7.21(월), 오후 3시, 44쪽–51쪽.

[32] Article 6 Para. 9 of the Broadcasting Law, http://elaw.klri.re.kr/kor_service/lawView.do?hseq=25243&lang=ENG.

[33] 김민환(책임)/한진만/윤영철/원용진/임영호/손영준, 방송의 공정성 심의에 대한 연구(방송통신심의위원회 용역보고서), 2008.

Table 6.1 All fairness review cases handled by KCSC for 6 years between May 2008 and before the Sewol Ferry incident in April 2014 and results

Date of decision	*Contents reviewed*	*Results*
July 16, 2008	MBC PD수첩 <미국산 쇠고기 광우병 관련> MBC PD Notes, American Beef, and Mad Cow Disease	Apology to the viewers[a]
November 26, 2008	YTN블랙투쟁/ YTN Black Suit Struggle	Apology to the viewers
February 18, 2009	KBS 1TV '[특별방송] 가는해 오는해 새 희망이 밝아온다' <제야의 종 관련 방송 왜곡>	Advisory
March 4, 2009	MBC 뉴스데스크 <미디어법 개정 관련>	Warning
March 4, 2009	MBC뉴스 후 <미디어법 개정 관련>	Apology to the viewers
January 27, 2010	MBC PD수첩 <4대강과 민생예산>	Advisory
January 5, 2011	KBS 추적 60분 <천안함 논쟁 관련>	Warning
March 10, 2011	KBS 추적 60분 <사업권 회수 논란, 4대강의 쟁점은?>	Advisory
July 7, 2011	MBC라디오 손에 잡히는 경제 홍기빈입니다 KBS라디오 박경철의 경제포커스 <유성기업 파업관련>	Advisory
July 7, 2011	MBC라디오 박혜진이 만난 사람들 <일제고사 거부 관련 해임/복직교사>	Caution
July 21, 2011	KBS-1TV 다큐멘터리 '전쟁과 군인' 2부작 <백선엽 다큐>	No issue
February 16, 2012	CBS라디오 시사자키 정관용입니다 <명진스님 인터뷰>	Caution
February 16, 2012	SBS김소원의 라디오전망대 <김용민 출연>	Caution
March 8, 2012	CBS김미화의 여러분 우석훈-선대인 출연	Caution
March 8, 2012	MBC생방송 오늘아침 <정봉주 전 의원 구속관련>	Opinion
April 5, 2012—January 9, 2014	KBS스페셜 '13억 대륙을 흔들다, 음악가 정율성'	First, deferral and then later, caution
July 5, 2012	MBC뉴스데스크 <민주당 난입>	Advisory

(continued)

Table 6.1 (continued)

Date of decision	*Contents reviewed*	*Results*
September 13, 2012	MBC뉴스데스크 <권재홍 “부상”보도>	No issue
September 13, 2012	KBS시사기획 창 <2012 노동자의 삶>	Advisory
March 21, 2013	MBC ‘특별대담-마유미의 삶, 김현희의 고백>	No Issue
July 11, 2013	채널A-김광현의 탕탕평평	Advisory
July 25, 2013	RTV <백년전쟁> ‘두얼굴의 이승만’, ‘프레이저 리포트-누가 한국경제를 성장시켰는가’	Warning and personnel actions
August 13, 2013	MBC황금어장 <안철수 편>	Advisory
September 12, 2013	KBS시청자데스크 <국정원 대선개입사건 보도 부실>	Advisory
November 21, 2013	KBS추적60분 <서울시공무원 간첩사건>	Warning (reversed in judicial review October 22, 2015)
December 18, 2013	TV조선 판 <정미홍, 박원순 종북 발언>	Opinion (3:2 subcommittee)
December 19, 2013	JTBC ‘뉴스9’ <통합진보당 해산청구>	Warning and personnel action
January 5, 2014 (소위)	TV조선 돌아온 저격수다 ‘박창신 신부 비하’ 방송	Advisory
January 23, 2014	CBS김현정의 뉴스쇼 <박창신 신부 인터뷰> - 국정원 댓글 논란, 연평도포격	Warning
March 20, 2014	MBC 뉴스데스크 <해고무효판결>	No issue
April 3, 2014	JTBC 뉴스큐브6 <유우성씨 인터뷰>	Warning and personnel action

Source collated from review transcripts uploaded on www.kcsc.or.kr

[a]The levels of discipline levied are, in an increasing order, Opinion, Advisory, Caution, Warning, Warning, and Apology to viewers. Starting with Caution, the broadcasters receive points off during the subsequent re-licensing review

What is notable is that KCSC's fairness review has been overwhelmingly on government views and programs. As shown in Table 6.1, almost all cases upon which KCSC conducted fairness review concerns the media coverage of contentious government policies such as concerning American beef import, appointment of publicly owned broadcasting executives, newspaper-broadcasting cross-ownership, a major river-bed renovation project, the investigation into the sinking of ROK Corvette Cheonan, police intervention in labor disputes at Yusung Industries, reinstatement of teachers fired for resisting the nationwide academic performance exam, and the arrest of former MP Jung Bong-Ju for defaming the then President Lee Myung-Bak. The only three exceptions also concern historical convictions of the ruling party on Baik Sun-Yup, Jung Yulsung, and Rhee Syngman.

The review was conducted by a nine-member board, six of whom are appointed by the ruling party and three by the opposition. The results of the reviews were striking. Interviews with Buddhist Priest Myungjin and Father Park Chang-sin were heavily disciplined for using strong language against the President, while the programs calling opposition party members the ultimate epithet in the country, "pro-North Korea", were lightly treated. Programs presenting views differing from the Ministry of Justice, the Prosecutors' Office, the Ministry of Education, and the Ministry of Agriculture were heavily disciplined even though other programs that simply relayed the government's positions verbatim were not disciplined at all. Even an MBC news item that criticized a court judgment cancelling the broadcaster's own dismissal of dissent employees received "no issue" despite its obvious self-interest in the news.

In this regulatory environment where broadcasters are receiving clearly lopsided signals from the regulator KCSC that does have the power to influence renewal of their broadcasting licenses, it was not surprising that MBC and KBS left out the bombshell eye witness Kim Hong-Kyung of the Coast Guards' fatal inaction during the crucial hours of the disaster.

After the Accident

Censorship of the critiques of the rescue efforts continued. Hong Ga-Hye was arrested for criticizing the Coast Guards for blocking volunteer divers from entering the ship.[34] Several people including the Japanese newspaper *Sankei*'s Seoul correspondent were arrested and/or indicted for questioning aloud the President's whereabouts for approximately 7 hours following the disaster,[35] a crucial element in evaluating the quality of the government's rescue command.

[34] http://www.koreatimesus.com/korean-court-acquits-woman-of-lying-about-ferry-rescue/.

[35] http://www.japantimes.co.jp/news/2015/12/17/national/crime-legal/seoul-verdict-looms-defamation-trial-ex-sankei-bureau-chief/.

Fatal Paternalism

"It is buildings which kill (people) in earthquakes", UNISDR's Twitter account cries.[36] Structures intended for shelter can be fatal. The same reversal is also true for Sewol. It was the ship built to keep people above water that drowned people in the Sewol ferry. It was the ship's steel walls that stood between the victims wearing the life vests and the free water that would have buoyed them up to the rescuers within a stone's throw in calm and warm water near the coast. All the waterproof technology that went into making the ship seaworthy and withstand water pressure now worked against the passengers, some of whom tried to break the windows in vain. One may as well say that Sewol would not have happened if the ship had been made of straw. In Estonia,[37] most people died due to hypothermia within 2 hours of entering the freezing Scandinavian water, in the aftermath of which experts around the world tried to design ships into "the best lifeboat(s)" that people will not have to leave in case of emergency.[38] That was not the case in Sewol: all the passengers had the life vests on and all they had to do was jump into the water. There is no single confirmed passenger that jumped into the water who was not saved.

This rather simple solution of jumping off the ship was famously accessible at multiple times.[39] Firstly, the infamous "Stay Still" order remained effective while the crew were abandoning the ship. The first distress call was made, not by the crew but by a passenger, a high-school student who called 911 on his mobile phone at 8:52 am when the ship had listed to about 30°. While the ship was listing further, the crew repeatedly issued "Stay Still" orders until 9:39 am when the last one was issued. The ship then listed to 53° by 9:35 am and further on to 64° by 9:54 am.[40] The Kakao Talk messages exchanged among the students

[36] February 6, 2016.

[37] https://en.wikipedia.org/wiki/MS_Estonia.

[38] Sturcke, James (March 6, 2007). "Herald of sea changes". *The Guardian*. Retrieved November 30, 2007.

[39] Except where expressly noted, all the data about the Sewol accident are from http://past.media.daum.net/sewolferry/timeline/.

[40] 합동수사본부 전문가 자문단, <여객선 세월호 침몰 사고 원인분석 결과 보고서>, 2014. 8.12, 4쪽.

show that there were many people who were standing by the "Stay Still" order as late as 10:17 am, at which one student famously wrote "They told us to stay still and no PA came out after that."[41] No "Abandon Ship" order was ever issued while the engine room crew abandoned the ship to climb onto a Coast Guard rescue boat at 9:38 am and it was not issued while the captain and one control room crew deserted the ship at around 9:48 am.[42] It was as only late as 10:15 am that one of the passenger service crew broadcast "Abandon Ship". All this means that there was at least about 1 hour during which people, cognizant of the danger, could have left the ship to their safety, had the crew not issued the "Stay Still" order or at least issued an "Abandon Ship" order as soon as possible or at least when the crew themselves were leaving the ship!

Some of the passengers disobeyed the "Stay Still" order and jumped into the water, and they were rescued around 9:50 am, that is after the captain and the entire crew had already left the ship. The ship that had been listing slowly began to list rapidly at 10:21 am and completely submerged at 10:31 am, leaving only its stern above water. Immediately before the final capsizing, about 40 passengers jumped off the ship over a few minutes starting at 10:18 am. This shows that an alert just a few minutes earlier would have given so many people a chance to protect themselves, had the "Stay Still" order not kept them from helping themselves out of the ship.

A second observation regards the mysterious refusal by the Coast Guard to enter the ship to either rescue or alert the passengers. The first help came as early as 9:23 am when the Coast Guard helicopter arrived at the scene. None of the rescue crew came into the ship to alert everyone to leave. The helicopter just rescued the people who had climbed up to the top (then the starboard side) of the ship. The Coast Guard rescue boat arrived at the scene at 9:38 am to pick up the engine room crew but none of the rescue personnel went inside the ship to alert the passengers to leave. A 116-ton Coast Guard ship actually reached the ferry at 9:48 am to pick up the captain and control room crew but none of the rescue personnel went inside the boat to alert the passengers to leave the ship. The Coast Guards could see the passengers inside the windows but

[41] MBN, "10시17분 마지막 카카오톡. . '기다려라 안내방송만. . .'", 2014.4.28, http://media.daum.net/breakingnews/newsview?newsid=20140428204606930.

[42] 합동수사본부 전문가 자문단 보고서 117–118쪽.

did not immediately start breaking the windows. In the end, for about 1 hour, none of the Coast Guards attempted to enter the ship to rescue anyone or alert the passengers to leave. It is only a painful irony that, while the rescue professionals were not entering the ship, at least two high-school students who had risen to the deck to jump overboard went back below decks to rescue their friends.[43]

The third point is about the post-submersion efforts. After the ship capsized and submerged, various efforts to enter the ship 30–40 meter underwater were hampered by currents and low visibility of 20–30 cm. In hindsight, there were clear chances to keep the ship in a better position, location, or shape for further rescue efforts before the ship went down that far. Life lines that the divers could follow to enter the ship underwater[44] were installed only three days after the accident on April 19.[45] Those lines could have been easily installed while the ship was afloat or immediately after the ship began submerging.[46] A diver even died trying to install an additional lifeline.[47] If officers at the scene dropped life lines into the ship so that the passengers could have held on while pulling themselves out of the ship, those lines could have been used later by divers on the way down. None of this happened.

Also, there were many ships near the scene that could have affected if not the depth, at least the rotation of the boat. By 1 am, April 18, a sea crane of 3,600 tons lifting power arrived, and another one of 8,000 tons lifting power had arrived by 1 am, April 20. The Sewol ferry was heavy, i.e., above 10,000 metric tons, but the combined lifting capacity of all the sea cranes by April 20 was above 20,000 metric tons.[48] Only one year after the sinking, the government is discussing recalling two of those sea cranes to lift up the Sewol ferry[49] clear of the surface. Certainly, there

[43] Hankyoreh Shinmun, "Yang On-Yoo Goes Back into the Ship to Rescue Her Classmate", 2014.6.17 http://www.hani.co.kr/arti/society/society_general/642883.html, Hankyoreh Shinmun, "To Chung Won-Suk, Who Went Back To Rescue His Classmate", 2014.11.10 http://www.hani.co.kr/arti/society/society_general/663864.html.

[44] http://media.daum.net/society/clusterview?newsId=20140420135704851&clusterId=1150909.

[45] http://news.mk.co.kr/newsRead.php?sc=60700149&year=2014&no=614414.

[46] http://dreamlives.tistory.com/964.

[47] http://www.edaily.co.kr/news/NewsRead.edy?SCD=JG31&newsid=01279206606084984&DCD=A00703&OutLnkChk=Y.

[48] http://www.munhwa.com/news/view.html?no=2014042401030243061002.

[49] http://nocutnews.co.kr/news/4402478.

was a talk of danger to possible survivors inside posed by the lifting plan but reducing the diving distance would have been safe and made easier the devastatingly slow rescue efforts that soon turned sadly into a salvage operation in a few days.[50]

This type of paralysis of common sense had featured in Korean disaster rescue scenes earlier. In December 2007, the Samsung Heavy Industries' sea crane collided with an oil tanker, causing an oil spill of the largest size that had ever hit the country.[51] The oil began to leak at 7:15 am from the tanker anchored near the coast and many villagers, watching the oil pouring out, suggested pulling barges by the side of the tanker to hold the oil leakage.[52] The affected area was famous for sand extraction. There were sand barges from 1,000 up to 3,000 tons. Just a few of them would have held most of the oil spilled. Also, there was a more instinctively commonsensical solution: plug the leaking hole. These common sense suggestions[53] were flatly ignored by the experts at the Coast Guard or instituted too late.[54] The holes were blocked only 48 hours later, during which

---

[50] http://bbs1.agora.media.daum.net/gaia/do/debate/read?bbsId=D003&articleId=5210913&RIGHT_DEBATE=R5.

[51] Wikipedia "2007 South Korea Oil Spill", https://en.wikipedia.org/wiki/2007_South_Korea_oil_spill.

[52] Youtube, "태안사태는 조작이다. 2부", https://www.youtube.com/watch?v=T72_Ki47j1c, Between 5:06 and 6:11.

[53] 둔재, "유조선에서 유출되는 원유 받기" 2008.01.13 블로그포스팅 http://m.blog.daum.net/imbg77/13656078 ("비록 풍랑이 거셌다고는 하지만 유조선이 침몰한 것도 아니고 멀쩡히 떠 있는 상태인데
조그만한 구멍 3개를 막는데 어떻게 이틀씩이나 걸렸는지."
풍랑이 높을 경우 대형 바지선을 유조선에 댔을 경우 2차 충동사고의 우려가 높으므로 유조선과 부딪치더라도 피해를 끼치지 않을 만큼의 소형 바지선으로 기름을 받은 후 인근의 대형바지선이나 또 다른 유조선으로 펌핑을 하면 가능하지 않았을까....)

[54] 주간동아, "2008.1.15, http://weekly.donga.com/docs/magazine/weekly/2008/01/14/200801140500009/200801140500009_1.html ("사고 발생 두 시간 만에 해경으로부터 연락을 받았습니다. 태안에 도착한 것은 사고 발생 5시간 만이었죠. 해경이 좀더 큰 배를 제공했더라면, 또는 선주 측이 조금 빨리 작업을 허락했더라면 어땠을지....") 염형철, "태안 기름유출사고와 기업의 사회적 책임", 태안신문 2008.4.24 ("당시 유조선의 5개 탱크 가운데 1·3·5번 탱크에 생긴 구멍을 첨단 타공(뚫린 구멍을 막는) 기술로 막은 시각은 사건 발생 48시간 뒤인 9일 오전 7시30분. 그러나 이미 1만500kl의 원유가 바다로 흘러나온 뒤였다.)
http://www.taeannews.co.kr/news/articleView.html?idxno=11765 (아쉬운 것은 유조선의 구멍을 48시간이나 방치하면서 흘러나오는 기름을 막지 못했다는 점이다. 유조선의 규모를 고려하지 않은 채 소형어선을 이용해 수리를 시도하다 두 차레나 실패하고 돌아오기를 연속한 탓이었다.")

more than 10,500 tons of oil were spilled.[55] This contrasts with what Chawle Singh, the tanker captain, did: he moved the oil around to put the ship off balance and raise the location of the holes relative to the oil.[56] If we can appreciate how much effort went into mopping up one gallon of oil off the sandy shore that the oil later hit,[57] it is of utmost regret that these simple remedies, no matter how partial, were not pursued.

Maybe the Coast Guards learned the lesson. When another oil spill took place seven years later in Busan on February 15, 2014, two Coast Guards immediately began the blocking operation (20 cm × 30 cm) and succeeded within 4 hours.[58] The oil spill damage was kept to a minimum. In the *Samsung Heavy Industries* Oil Spill, two large holes, 1.6 m × 20 cm and 1.2 m × 10 cm respectively, stopped pouring within 4.5 and 8.5 hours of the impact and a third small hole 30 cm × 3 cm poured out for 37 hours.[59]

A similar lapse of commonsense judgment took place in February 2003.[60] A subway train at a stop was set on fire by an arsonist and the fire spread to another train stopped in the opposite direction. As the dark smoke roared through both trains, most passengers of the first train exited the train to their safety. However, the driver of the second train,

---

[55] 경향신문, "손놓은 초동대처 하룻밤새 넋놓다. . 피해 왜 커졌나" 2007.12.09 http://news.khan.co.kr/kh_news/khan_art_view.html?artid=200712091915371&code=940202 ("유조선의 기름 유출 구멍을 신속하게 막지 못한 것도 피해 확산의 주원인이다. 당국은 유조선내 '바라스트(선박 균형장치)'로 기름을 응급 이송하면서 7일 낮 12시쯤부터 유출은 더 이상 없다고 발표했다. 하지만 기름은 계속 유출되다 48시간 만인 9일 오전 7시30분에서야 응급 봉쇄작업이 완료됐다.") ; 문화일보 "엉터리 예측 – 안일한 대응. . . '재앙' 키웠다", 2007.12.10 http://www.munhwa.com/news/view.html?no=20071210010305242190042 ("기름도 7일 낮 12시부터는 추가 유출이 없을 것이라고 발표했지만 9일 오전 7시30분쯤 1번 탱크에 난 길이 30cm, 폭 3cm의 구멍을 나무로 틀어막은 뒤에야 유출을 막았다.")

[56] 연합뉴스 2007-12-17 "유조선장 기지로 '6천800kl' 추가유출 막아"

[57] 김현근 블로그포스팅 2008.1.8. http://ryum2001.blog.me/140046738077, ("기름유출사건을 되돌아보면. . 인천앞바다에 지천으로 떠 있는 모래채취 바지선이라도 가져다가 쏟아지는 기름을 받았더라면. . . 웬만한 모래바지선 한척이 2–3천톤 정도의 크기이니 많은 양의 기름유출을 감소시켰을 것입니다.")

[58] 연합뉴스, 2014.2.15, "'부산 화물선 기름유출', 해경대원들 온몸 던져 막아", http://www.yonhapnews.co.kr/bulletin/2014/02/15/0200000000AKR20140215060300051.HTML.

[59] 정대진 Unpublished manuscript.

[60] Wikia.com, "대구지하철 화재 참사" http://ko.answers.wikia.com/wiki/%EB%8C%80%EA%B5%AC_%EC%A7%80%ED%95%98%EC%B2%A0_%ED%99%94%EC%9E%AC_%EC%B0%B8%EC%82%AC.

after the effort to move the train to safety proved futile, left the train as ordered by headquarters but took the master control key with him while the passengers were waiting for their next safety instructions. The trains were built so that if the master control key is pulled off, all the doors are immediately forced shut. As a result, 142 passengers were locked inside that train alone and burnt to death as the flames slowly consumed the whole body of the train.[61] With the master key off, all the lights were off in the dark underground except the emergency light and the passengers could not find the emergency switch that forces open the doors from within. Only the passengers in Cart No.4 were lucky: one of the passengers happened to be the chief of another train station experienced enough to locate and activate the emergency door switch in the dark.[62] Tellingly, the last safety instruction of the second train's driver immediately before he detrained was. "Please Stay Still."[63]

Why do these mass disasters happen? Lack of manual? Lack of discipline? Lack of investment in safety equipment? Probably, the Daegu Subway Operation Headquarters followed the manual when it ordered the driver to keep the key in his possession all the time, fearing that a runway train operated by wrong hands could wreak havoc, and the Coast Guard also followed the manual when it rejected the simple solutions proposed by the villagers, citing the possibility of explosion.[64] Accident manuals are provided to contain their consequences. Unanticipated accidents cause mass disaster. When accidents not included in the response manual happen, the only protection from a mass disaster is independent thinking drawing on common sense and the survival instinct.

Malcolm Gladwell in his bestseller *Outliers* analyzes the 1997 Korea Air Line accident in Guam. According to him, the cause of the accident

[61] 전북일보, "대구지하철참사, 기관사 '판단 잘못했다' 과실시인", 2003.2.21 http://www.jjan.kr/news/articleView.html?idxno=84577.

[62] 동아일보, "대구지하철방화참사 60여명 구한 금호역장 권춘섭씨, 2003.2.20 http://news.donga.com/3/all/20030219/7914510/1.

[63] 한겨레, "대구 지하철 참사 때도...'기다리라'는 방송", 2014.4.18 http://www.hani.co.kr/arti/ISSUE/124/633398.html.

[64] It is amazing that every stakeholder now wished that all the oil would have been better burned in an explosion than pollute the shore.

was the pilot's failure to abort the landing when his crew alerted him to the possibility of a collision. When the pilot did not abort, the co-pilot should have taken over and aborted the landing himself but he did not do that. Gladwell traces this behavior to Power Distance Index, a measure of how a sense of hierarchy deters people from challenging their superiors. The power distance index among pilots measured by scholars of cultural studies has a positive correlation with the number of airplane accidents.

Likewise, in Sewol, one can easily conjecture that a sense of obligation to conform to the hierarchy between the crew and the passengers suffocated a desire to speak up or act upon the crew's looming concerns or the passengers' genuine fear by jumping off the ship against the "Stay Still" order. Hierarchy between experts and lay people played an important role in Sewol: the passengers, facing obvious dangers, deferred to the judgments of experts. Of course, the real cause of the disaster was the incompetence and immorality of these experts but their failures could be easily corrected by the independent judgment of the passengers. That unfortunately had been suppressed by cultural forces. The impact of that hierarchy was maximized by the fact that most of the passengers were high-school students of Danwon High School (325 out of 477), out of whom only 75 were rescued (23%), while 91 were rescued out of 137 other passengers (67%).[65] Out of 14 teachers of that high school chaperoning the students, only 3 were rescued, eerily paralleling the rescue rate of the students, probably explained by testimonies that the teachers remained with the students until the last minute.[66] The darkest but most popular joke around Danwon High School was that "Only good kids died while naughty ones survived."[67]

It is not just active orders that suppress effective decision-making. The habit of respecting superiors' orders creates a dormant order that paralyzes independent thinking of people in inferior positions. The captain did not issue an "Abandon Ship" order just because the shipping company did not tell him to (the captain did not issue it even when the Coast

[65] https://ko.wikipedia.org/wiki/%EC%84%B8%EC%9B%94%ED%98%B8_%EC%B9%A8%EB%AA%B0_%EC%82%AC%EA%B3%A0.

[66] 조선일보, "'얘들아 올라와' 아래로 내려간 선생님들", 2014.5.21, http://premium.chosun.com/site/data/html_dir/2014/05/21/2014052100258.html.

[67] Private conversations.

Guard traffic control told him to at 9:22 am[68]); the crew did not issue an "Abandon Ship" order just because the captain did not tell them to; and the Coast Guards did not independently issue an "Abandon Ship" order upon arrival, one full hour before the submersion. The only independent thinking exercised against a superior's order was ironically by the captain of Coast Guard Ship 123, who refused to enter the ferry against a direct order from the superiors received at 9:47 am, fearing the capsizing, which did not happen until 10:21 am.[69]

Within a month after the accident, a high-school teacher working at the high school from which most student passengers came from opined, "Though not the primary cause, the school has to be blamed. Commonsense judgment must have dictated them to jump off. Children were handicapped in their ability to appreciate the danger." A newspaper commenting on the interview blamed the school for "adhering to injection-type of education focused on recitation, and thereby failing to nurture the students' ability to make autonomous judgments."[70]

At a symposium held about six months after the accident, a researcher at the provincial school board governing Danwon High school started a speech as follows: "Students are not to blame for following the experts' orders. However, we the educators must confront fundamental questions such as whether we coerced the students so deeply into unilateral obedience to authority and failed so utterly to nurture their independent and critical thinking that they just stayed still in the cabin obeying the public announcement, even facing the fatal situation."[71]

He continued, "The crew, who repeated broadcasting only the 'Stay Still' order waiting for the superior's direction, or the Coast Guards, who did not actively engage in rescue efforts just because there is no direction from above, present grave challenges. The government's disorganized rescue command is a probable manifestation of the same fundamental problem: the responsible officials were busy just waiting for and accommodating the whims of their superiors without any authentic sense of

---

68 JTBC, "9시23분 방송불가? 동영상이 증언한 선원들 거짓말", 2014.4.30 http://media.daum.net/breakingnews/newsview?newsid=20140430223813343.

69 경향신문, "해경청장 '선내진입' 지시했지만 경비정은 '경사심해' 불이행", 2014.5.18 http://news.khan.co.kr/kh_news/khan_art_view.html?code=940202&artid=201405181832291.

70 경향신문, "몸 마음 눈으로 세월호를 겪은 8인이 말하는 안전", 2014. 5.12 http://news.khan.co.kr/kh_news/khan_art_view.html?artid=201405112151095&code=940202.

71 장은주, "단 한명의 아이도 포기하지 않는 교육", 2014.9.30, 경기도교육연구원 심포지엄 < 위기의 한국교육 탈출구는 어디에>.

professional duty. *All of these point to the complete absence of the autonomous individual capable of independent and critical thinking without anyone's order or instruction* (emphasis added)."[72]

The point is not to ignore the experts or the manual, but not to force people to follow them blindly even when experts and manuals do not talk balk. Experts and manuals are given due authority by people when they present persuasive explanations for their instructions. What shall people do when there are no explanations for or no responses to question instructions like "Stay Still"? Authority is cyclical in certain cultures: experts and manuals must be followed without question because they are authoritative. Such authoritarian culture paralyzes the autonomous thinking that may have saved themselves and their friends.

CNN also touches on the same issue.[73] The coverage triggered heroic rebuttals[74] by sensitive authors calling the coverage "ethnic stereotyping" and it is surely incontestable that we cannot blame the victims and that "government structures" are mainly to blame. It is just that the mechanics of the disaster was so otherworldly (i.e., there were very clear opportunities for diversion that people even under worse governments could do well to avoid such disaster), and that people under good governments would have done badly under those lapses of judgement. Asking for better government is not an answer to the socially catastrophic nature of the disaster and begs questions on exactly what type of government could have prevented the fatal lapses described above. We should try to figure out what constitutes a good government, and I think that one of those qualifications are laws that promote and reward independent thinking, by abiding by freedom of speech and equality, among others. When we blame the hierarchy, we do not blame victims but the experts and officials who have forced them to accept their bureaucratic, manual-driven solutions. Stereotyping hinders learning about the causes of the phenomena. This inquiry calls for the exact opposite.

---

[72] Id.

[73] "South Korean cultural values played a role in passengers staying below deck as ferry sank", 4.18.2014, http://thelead.blogs.cnn.com/2014/04/18/south-korea-cultural-values-played-a-role-in-passengers-staying-below-deck-as-ferry-sank/.

[74] http://www.vice.com/read/stop-blaming-south-koreas-culture-for-last-weeks-ferry-disaster. http://time.com/75742/south-korea-ferry-asian-stereotypes-culture-blaming/.

Conclusion

People's independent thinking was held hostage by the laws, regulations, and indictments suppressing it. It showed through the captain who did not issue an "Abandon Ship" order just because the shipping company did not order him to; through the crew who did not alert the passengers to leave the ship just because the captain did not issue such order; and through the Coast Guards who did not broadcast or otherwise alert or guide the passengers to abandon ship just because there was no such order. In particular, laws that put dissident whistleblowers at risk of criminal indictment unless their concerns are packaged in collectivistic approval of "public interest" also not only nurtured disaster-prone corruption but also oppressed the independent individual.

The relationship between disaster relief and efficient communication was demonstrated in a study done on Twitter and how it contributed to obtaining the most accurate information for disaster relief. The researchers reviewed tweets RT'd in the wake of the February 2010 Chilean earthquake and found that Twitter users' responses worked as a filter that buoyed up more truthful tweets while letting unreliable tweets sink. For safety, as many people as possible should be able to speak to all others unhindered so that the free market of ideas operates to let truths trickle down, and the more urgent the situation is, the faster the free market should operate.

The government that forced into silence the worker who could have averted the fateful overloading of Sewol, and the hero rescuer who could have alerted the public of the incompetence of the professional rescuers, is ordering the people to "Stay Still", much like the captain who ordered the passengers to "Stay Still." If people obey like the high-school students did, they will be hit with another Sewol or more. Free speech and other institutions friendly to independent thinking of individuals must urgently come to the rescue.

References

Roos, W. (1998). Case comment, Netherlands: Copyright: Right to privacy and portrait right. *Entertainment Law Review, 9* (8), N146–N147.

Vaughan, D. (1996). *The challenger launch decision: Risky technology, culture, and deviance at NASA*. Chicago: University of Chicago Press.

Author Biography

Park Kyung-Sin is Korea University Law School Professor and one of the founders of Open Net Korea, and has organized several impact lawsuits at PSPD Law Center. He has written numerous articles on free speech, privacy, judicial reform, and so on. An alumnus of Harvard University (1992, Physics) and UCLA Law School (1995), licensed in California and Washington State, he is also the founding editor (2007) and the Editor-in-Chief of *Korea University Law Review*, available on Westlaw. He is also the founder of the Clinical Legal Education Center of Korea University (2009). He has served as a Commissioner of the Korean Communication Standards Commission (2011–2014), appointed by the President of ROK, and a Member of the Civilian Commission on Media, appointed by the National Assembly (2010).

CHAPTER 7

Disciplining High-School Students and Molding Their Subjectivity in South Korea: A Shift in Disciplinary Paradigm

Seungsook Moon

On April 16, 2014, a large ferry named Sewŏlho sank in the southwestern sea of the Korean Peninsula and left 304 passengers dead. A majority of these casualties were junior students of Tanwŏn High School who were bound to Jeju Island, a beautiful semi-tropical island with UNESCO-designated heritage sites, for their study trip. Comprising over 80% of the casualties, the deaths of these high-school students stirred up

Research for this paper is supported by Elinor Nims Brink Fund of Vassar College Faculty Research Grants awarded in the Fall of 2015. I thank Christina Cho for her valuable research assistance and Ms Hyun-jin Kim for helping me observe the daily lives of high-school students. I am also deeply thankful to Mr Rae-gŭn Pak and a group of mothers of deceased students whom I met when I visited Tanwŏn High School and the Association of Bereft Families on December 28, 2015.

S. Moon (✉)
Department of Sociology, Vassar College, Poughkeepsie, USA

J.-J. Suh and M. Kim (eds.), *Challenges of Modernization and Governance in South Korea*, DOI 10.1007/978-981-10-4023-8_7

intense and widespread indignation and discussion.[1] In particular, these high-school students were told to stay where they were while the ferry was sinking slowly over a day. Both domestic and foreign mass media reporting about this tragic disaster commented that young students were disciplined to be submissive and compliant and this contributed to their mass deaths. To what extent is this description accurate? If so, how are they disciplined in school? Is the disciplining of students the continuation of old authoritarianism that demanded deference to authorities without raising questions and visible resistance? Or is it based on a new model different from authoritarianism? To what extent was their compliant "habitus" (internalized disposition) generated by excessively rationalized guidance that parents and schools provide to make them high performers? This excessive guidance resonates with the transnational practices of so-called "overparenting" or "helicopter parenting" that have been observed in urban upper-middle-class families in South Korea, the United States, and elsewhere (Deresiewicz 2014). This chapter will explore these questions to examine how young people are disciplined in high school in South Korea.

First, I will discuss the general contour of high-school education in South Korea with the focus on the twenty-first century. I will pay attention to the role of high school as the Althusserian "ideological state apparatus" and highlight the official categories of high schools that reflect the state's instrumental approach to secondary education as the supplier of useful and functional human resources. Then, I will provide a vignette from students' daily lives on the ground at Tanwŏn High School to convey the predominance of college entrance to the category of "general high school," which represents a large majority of high schools in South Korea. Third, I will analyze how a large majority of high-school students are disciplined in South Korea. Here I will underscore the ongoing shift in the major disciplinary paradigm from the authoritarian and militaristic model to the psychological counseling model. In conclusion, I will assess the implications of this transformation in connection to the normative subjectivity of young people being molded.

[1]For example, see Han (2014), Mun (2015), Song (2014a, b), Yi (2014), and Yi, Tong-yŏn.

High-School Education in South Korea

Schools as an "Ideological State Apparatus" and Instrument to Train the Labor Force

Since the institutionalization of schooling for mass education in the late eighteenth century, modern states have employed formal education to mold their young people into productive and functional members (Ramirez and Boli 1987). In the case of South Korea, this general role of schooling has been accentuated by the urgency of rapid economic development and the Confucian emphasis on learning. As a country with relatively little natural resources, Korea had to rely on its human resources and adequate education was necessary for turning young people into diligent and productive members. The Confucian legacy of associating education with the highest social status (in the hereditary status system of the Chosŏn Dynasty) predisposed individual Koreans to be enthusiastic about formal schooling and academic degrees as a high-status symbol. The Korean state, which remained visibly authoritarian between 1948 and 1987, implemented an educational reform roughly every 10 years since 1954, for political reasons rather than educational reasons.[2] The developmental state in particular, swiftly expanded high-school (as well as middle-school) education during the 1960s and 1970s to supply a well-disciplined labor force to fuel rapid industrialization and urbanization.[3] It had also maintained its tight control over formal education as the ideological state apparatus (Althusser 1971), imposing uniform standards for school curricula and supervising the regulation of students'

[2]There have been seven educational reforms since 1954 and all were politically motivated except the sixth reform. The third reform implemented in 1968 was most explicitly political, marked by the proclamation of the National Education Charter (국민교육헌장). The seventh reform carried out in 2007 was the most sweeping one, as a result of the growing attention to globalization and international competence after South Korea became a member of the Organization of Economic Cooperation and Development (OECD) in 1996 (Ch'ŏn et al. 2013, 16).

[3]Towards the end of the colonial rule, less than 1% of the entire population in Korea received secondary education and higher. Although South Korea was one of the poorest countries in the world in the 1950s, Koreans showed strong enthusiasm for formal education. In 1957, more than 90% of the entire population attended primary school. By 1990, more than 90% of the population attended middle school and in 1999, over 90% of the population attended high school (Ch'ŏn et al. 2013, 13–14).

daily lives within schools and without (Hwang et al. 2013, preface). The tight control over school curricula was also coupled with the mandatory adoption of state-approved textbooks (Moon 1997). The primacy of political reasons over educational reasons in these education reforms suggests the extent to which the state is conscious of formal schooling as its ideological tool to mold the youth.

However, the state's tight control over high-school education has been differentiated between urban and rural areas. Most significantly, in 1974 the developmental state equalized high-school education by eliminating hierarchical ranking of schools and differentiated entrance examinations, and introducing a standardized test and a lottery system for high-school entrance. While this standardization (고교평준화) has been considered the single most salient feature of high-school education in South Korea, it has been a phenomenon largely limited to major cities. Initially, the standardization of high school was introduced to the metropolitan areas of Seoul and Pusan in 1974 and expanded to the major cities of Taegu, Inch'ŏn, and Kwangju in 1975. Since then the expansion of high-school standardization stagnated due to financial and other social reasons and it has vacillated between gradual expansion and temporary contraction over time. As of 2013, the standardized system was adopted in 31 urban areas, covering 54% of "general high schools," which have represented a majority of high schools and a majority of high-school students (Hwang et al. 2013, 17, 33). In terms of land mass, these 31 urban areas do not represent a majority, but in terms of population density, they do. This is because South Korea has become highly urbanized and as a result, an absolute majority of Koreans have resided in major urban areas. The extremely high population density in major cities is poignantly captured by the following statistics: approximately half of the 50 million Koreans reside in Seoul and the Kyŏnggi Province surrounding the capital city. Consequently, educational policy reforms or academic and public discourses on high-school education have revolved around how to mitigate problems of the standardized system and devise an alternative to it (Ch'ŏn et al. 2013; Hong 2013; Hwang et al. 2013; Ku et al. 2008; Daejeon Metropolitan Office of Education 2011).

The official categories of high schools that have existed in South Korea over time reflect how the state views high schools as a mechanism to train its labor force. Prior to the standardization, there were basically two types of high schools. One was "humanities high school" (인문고) that prepared students for college entrance tests and college education and the other was "vocational high school" (실업고) that prepared students for working in specific trades like farming, fishing, industrial manufacturing, and trading. Although a few additional categories of schools were established for specific training in arts, railroad operation, athletics, and broadcasting, the two major categories represented most high schools.[4] With standardization, these additional schools were consolidated into the single category of "special-purpose high school" (특수목적 고등학교). In 1974, eight high schools were designated as such. This category expanded to include a science high school (과학 고등학교) in 1986 and a foreign language high school (외국어 고등학교) in 1992. In 1998, Pusan Design High School (부산디자인 고등학교) was created as the first "specialization high school" (특성화 고등학교) and in 2002, high school for the gifted was incorporated into this category. These categories are not peculiar to South Korea and are fairly in line with the common dichotomy between college-bound education and vocational training without a college degree that we can observe in other societies. What can be somewhat peculiar is the broad context of high-school education characterized by standardization. Both special-purpose high schools and specialization high schools were designed to provide students with specialized knowledge or techniques that were missing in the standardized education provided by the general high school.[5] To mitigate the problem of uniform and rigid control over school curricula and administration, the democratizing state (in post-military rule South Korea) created the categories of "autonomous high school" (자율형 고등학교) and "self-reliant high school" (자립형 고등학교) in 1999 and 2002, respectively. Under the 2010 policy to diversify high schools, these various and rather confusing categories were streamlined into the following four: (1) general high schools (일반 고등학교), which prepare

---

[4]For example, in 1953, the first art high school (서울예술고등학교) was established and in 1967, a railroad high school (철도 고등학교) was founded in conjunction with the Bureau of Railroad. In 1973, an athletics high school (체육 고등학교) and a broadcast high school (방송 고등학교) were founded to supply workforce trained in these specific techniques (Hwang et al. 2013, 19).

[5]The fine distinction between the two is rather hierarchical. The special-purpose high school deals with students who are gifted or high performers in the fields of natural sciences, arts, and sports whereas the specialization high school deals with students who are not college bound because they are pursuing either vocational training or alternative schooling (mostly as a result of having problems with conventional schools or avoiding such schools for ideological reasons) (NLUEE 2012a, b).

students for college entrance tests and college education; (2) special-purpose high schools (특수목적 고등학교), designed to teach gifted or highly skilled students in natural sciences, foreign languages, arts, athletics, and industrial techniques; (3) specialization high schools (특성화 고등학교), which provide students with vocational training or education alternative to the general high school; and (4) autonomous high schools (자율형 고등학교), which are given more autonomy in school curricula and administration (Hwang et al. 2013, 19, 26, 30).[6] This concise survey tells us two important things about how the Korean state has approached high-school education. During the period of authoritarian rule, it displayed a blatantly instrumental view of education as a tool to produce skilled and useful workers and managers that the industrializing country required. This explicit nature is clearly mirrored not only in the dichotomy of college-bound training and vocational training, but also in the creation of "special-purpose high schools" and their enduring presence into the period of post-military rule. After formal democratization, there was pressure to pay attention to such formerly neglected issues as autonomy and choice for individual schools (as well as students and their parents).

A Gap Between Normative Ideals and Daily Practices of Schooling on the Ground: A Vignette from Tanwŏn High School

What is noteworthy about high-school education in South Korea in the twenty-first century is a conspicuous gap between the normative goals of high-school education and the daily experience of schooling among students (and teachers) on the ground. In principle, high-school education is designed to promote ideals of democracy, equality, global citizenship, human rights, and self-realization. In terms of more concrete educational objectives, it is also meant to strengthen "character education" (인성교육) by inculcating in students such moral values as honesty, credibility, forgiveness, responsibility, and consideration for others (Seoul Metropolitan Office of Education 2014). Daily practices

[6] In 2012, there were 2305 high schools and 1.92 million high-school students in South Korea as a whole. There were 1529 general high schools with 926,966 students. There were 499 specialization high schools with 223,068 students. There were 128 special-purpose high schools with 43,976 students. There were 142 autonomous high schools with 94,242 students (Hwang et al. 2013, 30).

of high schooling on the ground, however, are organized primarily around the specific and narrow goal of college entrance, more specifically the entrance to universities located in Seoul or prestigious universities located in other areas. This is particularly true with general high schools (일반고등학교), whose goal is to prepare students for college education. Yet the predominance of this goal in the daily lives of high-school students is by no means limited to this category of high school. Among autonomous high schools, enjoying more leeway to decide curricular content and the cost of tuition, some schools have virtually become high-end schools specializing in teaching subjects crucial to college entrance examination (Hwang et al. 2013, 18). Special-purpose high schools which can attract students who are gifted or strong in natural sciences, foreign languages, arts, and sports have virtually become elite high schools, which send a large number of their students to prestigious universities in Korea and send a handful of their students to prestigious universities in the USA and elsewhere in the world. This unintended development has contributed to the weakening of general high schools and even their "crisis" of becoming a "wasteland" (Daejeon Metropolitan Office of Education 2011; Hong 2013).[7]

The collective zeal for entering prestigious universities and increasingly just any university located in Seoul (due to growing competition as a result of popularization of college education) is so intense and excessive that families with high-school students not only invest a huge

[7]Discourse analysis of mass media (both printed and online) covering the high-school diversification policy shows the following results. First, the public shared a concern about the intensification of competition resulting from diversification because this has revived hierarchy among different types of high schools. Second, in line with this concern, the public perceived the autonomous (private) high school negatively because this type of school has become an "aristocratic school" in the midst of the competition over the college entrance examination. Third, the autonomous (public) high school was associated with the fostering of talented human resources and higher academic performance, in comparison with the general high school lacking such merits. Fourth, residential high schools established in rural and other marginalized areas were seen to ease educational disparity among regions and improve educational environments in marginalized areas. Fifth, the Meister high school, a subcategory in the special-purpose high school, was perceived to be positive in fostering talented and competitive human resources in collaboration with industry and business. Sixth, the public perceived that the general high school was excluded from the diversification policy and faced with the problem of lowering of its quality (Hwang et al. 2013, p. v).

portion of their income to pay for commercial cram schools and private tutors, but also organize family lives around the class schedules of their children in high schools.[8] In fact, the national competition for getting higher scores in the college entrance examination begins far before children enter a high school. Among affluent upper-middle-class families in major urban areas, it begins when children enter a grade school or even a kindergarten.[9] For lower-class families, there can be two scenarios. If students and their parents have given up their faith in education as an effective means to achieve upward mobility due to actual and potential obstacles on their way, they are not motivated to join this race.[10] If they still keep a hope for upward mobility through higher education, they

---

[8] These commercial educational services available outside schools are called "private education." According to OECD statistics and Korean government statistics, in 2014, each Korean spent on average W 242,000 (roughly US $ 242) per month for such private education and each student (from grade school to high school) spent on average 5.8 hours per week for such education. Almost 70 percent of students in primary and secondary education used private educational services. (http://blog.daum.net/educationandsociety/812, accessed on 7/16/2016). Among the OECD member countries, South Korea has ranked number one in terms of private educational expenses. (http://www.ggilbo.com/news/articleView.html?idxno=292201, accessed on 7/16/2016).

[9] This is very similar to the helicopter parenting among the affluent and well-educated upper-middle-class families in the USA (Deresiewicz 2014).

[10] There has been growing erosion of the popular faith in education in South Korea, as it has witnessed growing inequality among different socioeconomic groups, especially since the 1997 financial crisis. When I visited a private (all women's) general high school located in the southwestern part of Seoul in late December 2015 to get a sense of daily lives of high-school students in Seoul, I observed a sizable minority of students who were not engaged during their classes in several classrooms I visited. Some of them were sleeping, others were solving test questions (to improve their skill at taking tests efficiently). There were even students who were applying and fixing their makeup. My visit to this school took place toward the end of the fall semester after students had taken their final examinations. Hence, the semester was virtually over and teachers were spending their class time showing movies and documentaries, administering games, or making students present a mock trial. When I asked teachers if such disengaged student behavior stemmed from the interval period, I was told that a sizable minority of students were always disengaged from studying even during the regular semester. The teachers do not chastise them as long as they do not overtly disrupt class because efforts to discipline them during class can interfere with the smooth process of teaching and learning. This school is located in a solidly middle-class area of Seoul where families have shown enthusiasm for education. Therefore, my observation of academic disengagement was particularly striking because it implied a rapid speed of erosion of the conventional faith in education among young students even in a middle-class context.

make their best efforts by using whatever limited resources available to them. This type of aspirational behavior was noticeable at Tanwŏn High School, 246 of whose students were drowned during the Sewŏlho disaster. This is a general high school located in the provincial city of Ansan, a largely working-class area with an influx of a large number of foreign migrant workers, in Kyŏnggi Province.[11] It was as recently as 2013 that this city, along with the cities of Kwangmyŏng and Ŭijŏngbu in the same province, was integrated into the standardized high-school areas (Hwang et al. 2013, 26).

When I visited Tanwŏn High School in late December 2015,[12] I encountered multiple signs of zealous dedication to college entrance in the 10 classrooms of the (entire) junior class that were preserved as memorial sites. Each classroom has 30 some desks and chairs used by individual students. On each desk of those who died, there are photos of a student, flowers, small plants, cards, and their favorite snacks brought by families and friends. Their desks and chairs were also decorated with yellow ribbons or yellow knitted covers, symbolizing unfulfilled wishes for their safe return. The desks of surviving students are tidy but nothing was displayed on them. These classrooms are minimally furnished with a television monitor, a large blackboard, a few bulletin boards, small wooden lockers for students' belongings, a few electric fans on a side wall, a calendar, and a mirror. There is also a compartment for keeping cleaning equipment. The blackboard is filled with written messages and drawings done by grieving students, expressing their feelings of love, sorrow, and guilt, their good wishes, and an articulation of memories. In the midst of these commemorative articles, there are common and unmistakable paraphernalia reflecting the centrality of college entrance to the daily lives of high-school juniors. First, a calendar clearly shows the countdown to D-day, the college entrance examination day, and each month shows a map of each province and the greater Seoul

[11]The socioeconomic marginality of these students and their families is poignantly captured by the fact that there was not a single parent who was a lawyer or a doctor among the 246 students drowned in the ferry disaster.

[12]My visit was made possible by Mr Rae-gŭn Park who has been working with the parents of deceased students. I also met some of these parents when I visited the collective memorial altars for victims of the ferry disaster. For their own collective support and healing, these mothers and fathers created handicraft workshops in a makeshift building where they gather to learn and manufacture small things and have also organized public protests.

area with markers indicating locations of four-year universities and two-year colleges. Second, bulletin boards on the back wall display posters, announcements, and guidelines related to the college entrance examination. For example, a poster lists individual students' names in a given class, names of the university they want to attend, majors they aspire to study, and jobs they aspire to have. If a university is located in Seoul, this is clearly specified in parenthesis, implying its significance. The most striking among materials on these bulletin boards in each classroom are noticeable maps of the nation, its individual provinces, the capital city of Seoul and the surrounding Kyŏnggi Province with dots indicating the locations of four-year universities and two-year colleges. These maps also provide students with clearly organized lists of names of universities and colleges in each province and the greater capital region. Other examples include posters advising students to exercise and stay healthy because "physical stamina is a source of competitiveness" (체력이 경쟁력), a printout of guidelines for "self-guided study" (자기주도학습) after formal classes end, and the yearly schedule for mock college entrance examinations with very detailed contents of each subject to be tested. The common daily timetable for students posted on these bulletin boards (in each individual classroom) shows that the school begins at 8:00 am and ends at 4:00 pm; students have four classes in the morning, a lunch break, and three more classes in the afternoon. After cleaning their classroom and having a closing homeroom meeting with their designated class teacher, students have two more classes until a dinner break which lasts from 6:20 to 7:10 pm. After dinner, they stay for self-guided study until 10:00 pm. This vignette of students' daily lives at Tanwŏn High School helps us understand the context of discipline to which high-school students are subject.

High-School Discipline in South Korea: "School Rules" (학칙) and "Regulation of Student's Life" (학생생활규정)

As I discussed elsewhere (Moon 2005, ch. 2, ch. 3, and ch. 4), primary and secondary schools in South Korea have for decades used a wide array of authoritarian practices to discipline their students. In particular, junior high and high schools employed militaristic practices to exercise close surveillance over students' appearance and behavior, and thereby to shape their habitus. Such militaristic practices included not only the use of uniform, but also of strict and meticulous codes for school bags, hair style, socks, shoes, and even underwear "appropriate" for students. Those practices also included rigid and detailed codes of conduct "appropriate" for students.[13] Prior to the abolition of corporeal punishment in 2010, violation of these codes subjected students to corporeal punishment and collective punishment to inculcate obedience and collective orientation in students. From a comparative historical perspective, however, South Korea was not unique in borrowing such authoritarian discipline from the modern military. As Michel Foucault poignantly argued, the modern military provided ruling elites with the archetype of "disciplinary power" to mold the individual and collective bodies of the ruled and transform them into the useful and docile subjects (Foucault 1979).[14] What may be rather peculiar to South Korea is the persistence of such militaristic disciplinary practices even after its formal democratization. As discussed below, a majority of these practices have outlived the period of military authoritarian rule (1961–1987). A significant shift in this militaristic discipline began at least formally with the elimination of corporeal punishment and the introduction of "students' human rights" (학생인권) paradigm in 2010 (Kyŏnghyang Newspaper 2010).

While high-school curricula in South Korea are designed to transmit standardized knowledge and develop cognitive skills (to be tested in the college entrance exam), "school rules" and "the regulation of student's daily living" are designed to control and guide student's appearance and behavior and thereby shape their habitus and subjectivity. These categories of regulation involve bodily discipline. In normative discourses on schooling in South Korea, discipline refers not only to the restrictive control of students' behavior to maintain a safe and orderly educational environment, but also to students' autonomous ability to regulate their own behavior.[15] However, quotidian practices of disciplining high-school

---

[13] My personal conversations (in November 2015) with a few teachers at general high schools in the Seoul area reveal that the model of a desirable or ideal student remains similar to that of the 1960s and the 1970s: the student who obeys her/his teacher (선생님 말씀 잘 듣는 학생).

[14] For example, Britain, commonly known for its parliamentary democracy, had its secondary schools employing some such militaristic practices in the early- and mid-twentieth century (Cookson et al. 1985; Deslandes 2005; Kalton 1966).

[15] It is through "discipline" (훈육) that students develop their reason and autonomy (Ko 2014, 9). Discipline style refers to a whole set of regulations that guide student behavior to maintain order in class and school life (Ko 2014, 6).

students on the ground show a heavy emphasis on external coercive regulations at the expense of students' autonomy and participation. By critically reading the discourse of high-school reform, focusing on school rules and the regulations of students' daily living in the twenty-first century, I will argue that the old authoritarian discipline is being replaced with a new paradigm of discipline emphasizing students' autonomy (자율), human rights (인권), and participation (참여) (Kyŏnggi Provincial Office of Education 2015; South Kyŏngsang Provincial Office of Education 2013). However, this new paradigm may not be a thoroughly democratic progress to be celebrated (or at least accepted), but rather the spread of Foucauldian disciplinary power that produces docile and useful subjects in a more civilized manner. This new disciplinary paradigm also shows Weberian rationalization marked by scientific thinking, instrumental thinking, and the systematic arrangement of human behavior.

In 2003, the discourse of primary-, middle-, and high-school discipline reform was galvanized by widespread problems of harassment and violence among students and increasing violence against teachers committed by students and parents in school. These cases of violence have been not only verbal and physical, but also virtual in cyber space. The most common form of harassment among students has been exclusion from peer groups.[16] Verbal abuse (e.g., cursing and shouting) has been the most widespread problem among both male and female students. According to a national survey, 73.5% of students (with similar responses for both genders) replied that they had experienced verbal abuse and 35.8% of students replied that they had experienced harassment in the form of exclusion (Kyŏnggi Provincial Office of Education 2015, 43). Cyber violence was committed through e-mail, cell phone text messages, and other internet communication sites. Gang groups commit violence and extort money and other material resources in schools, and these groups are linked to those at higher-level schools or adult criminal networks. It is not entirely clear if these various forms of violence have actually increased or if public awareness of them has increased and as a result more incidents of violence have been reported. It is likely that

[16] The increase in school violence among students has alarmed teachers, parents, and educational administrators. A specific slang word "great exclusion" (왕따) has been commonly used to refer to violent behavior, ranging from collective beating, cruel and humiliating treatments, and extraction of money and goods.

both have contributed to the emergence of the discourse of school discipline reform.

Teachers and educational administrators attributed these problems of violence and harassment to uniform and authoritarian regulation of students' daily lives. They perceived such rigid regulation as a source of students' discontent, which in turn made them violent. In searching for fundamental solutions and preventive measures, those adults have adopted the rhetoric of "student-centered" regulation of students' lives (as opposed to "school and teacher-centered regulation"). The Ministry of Education (MoE) recommended all municipal and provincial offices of education to pay attention to "students' human rights" (학생인권) and "students' autonomy" (학생의 자율) when these offices reform how they regulate student's daily lives. The MoE instructed schools to ameliorate specific clauses in existing regulations, especially those potentially violating student's human rights, through discussion with students and their parents. It also asked schools to avoid military-style punishment, including collective punishment, corporeal punishment, the investigation of personal belongings, and the public cutting of students' hair with scissors as a punishment for violating school rules (Yi, C.-Y. 2003). These MoE guidelines reveal the pervasiveness and persistence of authoritarian and militaristic practices of disciplining students in the first decade of the twenty-first century.

The emerging discourse of school discipline reform was amplified by an announcement to eliminate corporeal punishment in schools located in Seoul. In 2010, Kwak, No-hyŏn, a "progressive" Superintendent of the Metropolitan Seoul Office of Education (교육감), announced the thorough prohibition of corporeal punishment in the metropolitan Seoul area. To ensure the efficacy of this prohibition, he asked students and teachers to foster a new school culture (Yonhap News, November 16, 2010). This drastic change did invite opposition and resistance, particularly from the conservative Federation of Teachers' Associations, in the name of protecting teachers' authority to educate and guide students from being further eroded (Korean Federation of Teachers' Associations 2012). Despite this type of resistance, the abolition of corporeal punishment spread to the rest of the country and set the stage for a shift in the way high-school students are disciplined in schools and without. It enabled the rise of a new disciplinary paradigm, characterized by psychological counseling as both treatment and prevention under the rubric of "students' human rights."

In this discussion of the emergence of a disciplinary paradigm shift, it is important to recognize that students themselves also contributed to the ongoing transition from the militaristic model of school discipline to the human-rights model. For example, in August 2000, students collected signatures on the street and on the internet to loosen the strict regulations of their hair styles. Related petitions flooded the government agency for citizens' petitions. In April 2006, a group of middle-school students in Seoul protested against the excessive regulation of their hair styles. While school authorities intended to admonish leaders of the protest, a human rights organization opposed such potential punishment. This and other similar incidents triggered the Kyŏnggi Province Office of Education to instruct schools to listen to the viewpoints of students, parents, and teachers when they regulate students' daily lives and reform how they do so. In the 2000s, there have been publicized incidents of conflict between students and school authorities regarding the school rules. It appears that students have been growingly aware of their human rights and formed organizations to promote them. Some model schools were designated to experiment with students' participation in the making and remaking of school rules (Pak, C.-K. 2006).

There have been also certain efforts and experiments for promoting students' autonomy and participation in the classroom. For example, the seventh educational reform announced in 2007 introduced the new component of curricula differentiation, emphasizing academic competence and student choice in curricula. This reflects an assessment of high-school education in South Korea from a comparative perspective. In a comparative study of high-school education in the USA, Britain, Germany, China, Japan, and Australia, Korean scholars of education indicate that high-school education in South Korea has been marked by quantitative growth and expansion of equality to supply human resources necessary for national development. While this standardization has led to the reduction of academic disparity among schools, it has increased the emphasis on rote memorization to secure high scores in the college entrance examination. They argue that now is the time to enhance the quality of education and strive for excellence (Ku et al. 2008, 43). Other scholars of education argue that standardized or equalized education was suitable to a mechanized industrial model of production, characterized by small variations and large quantities. It is not compatible with the postindustrial society based on specialized knowledge marked by value-added production (Hwang et al. 2013). Although curricular differentiation can be beneficial for students, the persistence of the narrow and overarching goal of college entrance, however, undermines its potential.

Another example of loosening the uniform and tight control over high schools is "the subject-based classroom system" (교과교실제) experimented nationally since 2010. This system is an example of pedagogical experiments to enhance the quality of secondary education. It was adopted as a "class teaching innovation project" and considered to be a "major component" of preparing for the future of secondary education in South Korea (Ch'ŏn et al. 2013, 1). In particular, the system was introduced to recognize academic differences among individual students and provide them with classes at various levels. In 2009, this system was introduced as a test case in a few schools and in 2010, it was adopted widely. This was a relatively bold move, given that for six decades, the category of general high schools, geared to the college entrance examination, has maintained the fixed classroom system (학급교실제) with a "designated class teacher" (담임교사); this teacher is responsible for a fixed class of students who stay in the same classroom for an entire academic year and teachers of different subjects come to the classroom to teach the class. Designated class teachers are often those who teach major subjects considered to be important in the college entrance examination. The class teacher interacts with her or his students tied to a fixed classroom during her or his subject class, home room meetings, and extra classes after formal class hours (Ko 2014). It is not difficult to associate this system of a fixed classroom with unfree laborers or peasants tied to land or a workshop without the freedom of mobility. Although students are not like such indentured workers, this normalized practice does not go well with the new ideals of student autonomy, participation, and their human rights. It is worth noting that administrators and teachers were concerned about the potential problem of controlling mobile students under the experimental system. While an assessment report on the implementation of the subject-based classroom system conveys that this anxiety is not well founded, its implementation had only limited success because of the lack of classroom space to run it smoothly.[17]

[17] Research evaluating the implementation of the subject-based classroom system reports the following findings. The fundamental condition in Korean education is "equalization of high-school education" and this has generated the dilemma of how to deal with students with different levels of academic proficiency. However, its success depends on the following

The persistence of school rules and regulatory guidelines for students' daily lives in the midst of these changes suggests that they constitute the institutional mechanism of molding students through discipline within classrooms and without. These rules and regulations target a range of behavior perceived to be disorderly, harmful, and violent. The definitions of these adjectives can vary, depending on a society's moral values and normative standards. Some rules are clearly reasonable to most people but other rules are ambiguous in their objectives but curiously detailed. An empirical study of 41 humanities (later recategorized as general) high schools sampled nationwide in South Korea shows common examples of reasonable disciplinary practices in the late 1990s: disruptive or unfocused behavior in class and drinking and smoking in school. To deal with disruptive students in class, 71.1% of teachers surveyed replied that they made deliberate efforts to correct such students and 28.4% replied that they would give a mild warning and move on. To deal with such delinquent behavior as drinking and smoking, 14.7% of teachers surveyed used the heaviest punishment such as community service, social service, or special education; 48.7% of teachers used mild punishment such as school service; and 35.5% used chastising and ordering students to write a statement of regret. Based on studies outside Korea and inside, this empirical study points out that autonomous self-discipline is much more desirable than external and coercive discipline and student participation is more desirable than one-way order and control. (Han, T.-D. 2000). In terms of regulation of students' daily lives, it is worthwhile to note that these meticulous rules pertain to the appearance and behavior of students, including their hair and clothes. I argue that such meticulous regulation of appearance and behavior echoed both secular and religious regulation of the individual human body in pre-modern society with

---

Footnote 17 (continued)

conditions. The subject-based classroom system requires one classroom for one teacher, which means more space for classrooms. It also requires the long-term tenure of a teacher in a school that allows them to develop their own classroom with their own pedagogical methods and style. Some grade schools have this environment (because of the decreasing number of students) but it is impossible for public schools whose teachers mandatorily rotate amongst schools. In terms of class management, this new system can allow classes differentiated by the level of academic proficiency, addressing the problem of equalization and lowering the quality of learning among students with different levels of academic proficiency. In terms of regulating students' daily lives, the new system is not necessarily undermining necessary discipline.

hereditary status distinction based on caste and sex. Such bodily discipline is an age-old societal mechanism to generate docile and compliant habitus that accepts hereditary status distinction as being just and natural. The effect of detailed corporeal discipline on high-school students would be docile and compliant habitus in their individual selves.

In 2011, school rules in the nation were significantly modified by the Primary, Middle, and High School Education Ordinance Reform. Its salient features included the emphasis on students' human rights and the replacement of corporeal punishment with school or community service, completion of special education, and suspension. For example, the *Living Human Rights Education Basic Plan* published in 2015 by Kyŏnggi Provincial Office of Education, where Tanwŏn High School is located, conveys two points noticeably. One is its attention to democratic civil rights, including due process and protecting the civil rights of students in every step of the disciplinary process. The other is the heavy reliance on psychological counseling as treatment, prevention, and "healing" as a key word in the discussion of the rights of students and teachers (Kyŏnggi Province Office of Education 2015, 18, 25). This approach echoes the "well-being" (웰빙) trend which was in fashion during the first decade of the twenty-first century (in the aftermath of the 1997 financial crisis). These salient features of the change reflect an emergent shift, at least in principle, in the goal of disciplinary punishment (징계) from retribution to rehabilitation.

According to a survey of 148 high schools in the metropolitan Pusan area around the time when the 2011 reform was announced, the old authoritarian and militaristic school rules overlooked students' human rights. School rules in practice commonly included the screening of personal letters, prohibition of participation in various political meetings, and excessive restrictions on hair and clothes. An activist staff member in the Human Rights Education Center pointed out "because the central objective of school rules has been the control over students, they have violated human rights that the constitution has upheld." For example, a majority of high schools meticulously regulate their students' bodily appearance, just as a traditional religion has done for individual followers. School rules on hair style and clothes are often strikingly detailed about the length of hair, types of hairpins and accessories used, and the color or type of socks, stockings, and even underwear. Such details coexist with ambiguous or certain implicit assumptions; school rules dictate that students should wear "orderly and clean" (단정한) hair and "student-like" (학생다운) shoes. This type of regulation reminds us of common practices in traditional society with hereditary status distinction. In order to maintain inequality and mark status distinction, such a society regulated the clothes and hair of members of different social statuses. Some schools even have separate entrances for teachers and students; a central entrance is reserved for teachers and students should use side entrances. What is significant about this regulatory mechanism over the individual body is that it effectively produces and maintains internalized acceptance of authority. When individuals do not have control over the appearance of their own bodies, the physical base of their existence as living organisms, they cannot be independent and autonomous entities and they are reduced to being docile and useful components of a society or an organization.[18]

There were other anachronistic examples of school rules that produces conformity to and acceptance of authorities. Many schools limited leadership positions of class head and officer of students associations to students with high academic performance. There existed an array of residual restrictions on the development of students as autonomous citizens of a democratic society who could enjoy freedom of thought, conscience, and expression. Many schools prohibited student participation in voluntary associations outside school and required permission from the school president for such participation. A school president could disband a students' association in wartime and other emergency situations. Many schools also restricted students' involvement in political activities. Students who "secretly procure unwholesome texts, read, write, post, or distribute them" are subject to expulsion from school without even defining such texts. Teachers actually open letters sent to their students to examine their contents. Some schools even suspended or expelled students who refused certain classes on specific religion or patriotism. In sum, such basic civil rights as freedom of thought, religion, and expression, and such basic political rights as the freedom of assembly and the right to run for public office have been denied or curtailed in contradiction to the Constitution (Association of Parents for Genuine Education 2011; Chŏng and Pak 2011).

[18]See Association of Parents for Genuine Education (2011) and Chŏng and Park (2011).

Students' own perceptions of their teachers' interaction style and school discipline allow us to see glimpses of how the aforementioned education reform has affected the lives of students on the ground. According to a survey of 409 high-school students (conducted between late 2014 and early 2015), roughly 40% of students viewed their teachers "authoritarian," who interacted with them minimally and dictated them what to do without much discussion or explanation. Roughly 27% of students viewed their teachers "laissez faire," who were not interested in guiding them and interacting with them. Approximately 26% of the students viewed their teachers "accommodating," who were authoritarian but interacted with them a good deal. Interestingly, the authors of this study assert that this style of teacher was considered the most desirable for the school. About 7% of students perceived their teachers "affectionate," who were highly engaged with them without being authoritarian. The same survey also shows that approximately 50% of the students reply that the difficulty in school life includes their classes that are not interesting but taught only for college entrance tests, the excessive amount of work assigned for their classes, and the difficult content of their classes. About 23% of 409 students replied that difficulties in their school lives stemmed from strict school regulation. (Sim et al. 2015). This survey shows that a majority of teachers remain authoritarian in their mode of interaction with students and a sizeable minority of students is troubled by strict regulation of their lives in school.

Another survey suggests that there can be local variations in students' perception of their teachers' mode of interaction with them. According to a survey of 500 students at three general high schools in Ulsan Metropolitan area, students perceived their designated class teachers' disciplinary style as supportive (40.8%), dictatorial (31.6%), democratic (22.5%), or compromising and laissez faire (5.1%). These groups of high-school students included in the survey trusted their teachers more when they perceived their disciplinary style as democratic, supportive, or compromising (as opposed to dictatorial or laissez faire). The Ulsan area has been developed around Hyundai shipbuilding and automobile manufacturing industries and is considered socially and politically conservative. This conservatism may account for students' perception of lower authoritarianism in their teachers' attitudes. Yet, these students appreciate supportive, considerate, encouraging attitudes among their teachers, like to participate in class, be responsible, and to have choices, and dislike authoritarian demand, strict control and coercion, their teachers' lack of interest and expectation, and passivity from their teachers (Ko 2014).

A Shift in Disciplinary Paradigm: Subjectivities Molded by the Old Militaristic Practices and the New Practices of Human Rights and Counseling

The 2011 Education Ordinance Reform has brought formal change in the paradigm of discipline used in high schools but disciplinary practices on the ground experienced by high-school students suggest the persistence of old authoritarian practices in the process of spreading students' human rights paradigm characterized by psychological counseling. Although it is too early to assess its effects on the daily lives of students and the molding of their subjectivity in the long run, there are some important implications of this ongoing shift. First, in connection to the abolition of corporeal punishment, the emphasis on students' human rights and the growing significance of counseling can be seen as being indicative of a civilizing process (*à la* Norbert Elias) of reducing physical violence, brutality, and cruelty (Elias 2000). Although this civilizing process would be a positive development, it does not necessarily or automatically lead to an increase in individual liberty and autonomy. Second, as Foucault argued, this shift is accompanied by the pastoral power that paternalistically guides and takes care of the ruled. This can be seen as pacification and sophistication of the power that rules in modern and postmodern society. The empirical example of this can be seen in the specific form of disciplinary punishment for students who commit violence against other students. The goal of such punishment is professedly "the restoration of positive relationships between students," rather than retributive justice of punishment. Moreover, in the absence of corporeal punishment, words have become the primary medium of guiding and disciplining students. It is in this context that counseling emerges as a core technique of preventing delinquency, violence, and other disciplinary problems in advance. Designated class teachers are to improve their ability to counsel and a professional counselor is to be hired in each school.

The ascendancy of counseling is more visible in "special schools" dealing with students with problems and who therefore cannot attend regular schools. To discipline such students, Wee centers were founded in each educational district and a "Wee class" is offered (Kyŏnggi Provincial Office of Education 2015, 81). The Kyŏnggi Province has 25 Education Support Bureau and each of them established one Wee Center, generally staffed with two professional counseling teachers, two to three professional counselors, one clinical psychologist, and one social worker (Ibid. 92). These centers have used various forms of counseling, ranging from telephone counseling and face-to-face individual counseling to internet counseling. Each province has a Wee School and the first of these schools opened in October 2013. These Wee schools are special residential schools for "high-risk" students who need to get alternative counseling education in addition to regular school curricula. For example, "Hope School" provides education and counseling for unwed mothers and fathers and other students who cannot adjust to regular school life and have dropped out.

What sorts of habitus and subjectivity are being constructed or intended to be produced under the new disciplinary paradigm? The pastoral and therapeutic power of psychological counseling coupled with the rhetoric of human rights and the autonomy of students would mold habitus and subjectivity rather different from those produced by the old authoritarian paradigm of discipline. Because it is coercive and blatantly intrusive, the old paradigm of discipline allows students to develop resistance easily and develop an interior in their subjectivity that the coercive power does not reach. To the extent which coercion is used, students under such discipline do not easily trust the violent power but fear it. In contrast, the new paradigm of discipline with pastoral power of counseling and the scientific tools of personality tests would leave much less room for resistance and an interior not monitored by the scientific technique of meticulously categorizing individual psychology. Gentle pastoral guidance of counseling or scientific categorization and analysis is more difficult to resist because it appears to be caring and impersonal.

However, it is important to recognize that the difference between the two paradigms above is an ideal-type discussion and the reality in contemporary Korea can be more complex and even contradictory than the neat dichotomy can explain. Here we need to pay attention to the specific conditions of students' daily lives in school that I try to capture with the vignette from Tanwŏn High School. If the common daily schedule of those junior students in the 10 commemorative classrooms is representative of lives of students in general high schools in South Korea, their lives are engulfed by studying for the college entrance examination in school, at least from 8:00 am to 10:00 pm, excluding transportation time between home and school. They go home only to sleep. For those students who subscribe to the ideology of upward mobility through studying hard, they have to learn to manage their time and lives efficiently. This means that they have to internalize instrumental thinking and organize their behavior systematically, which Max Weber theorized as elements of rationalization. When individuals acquire this rational orientation and it becomes their habitus, they are very effective in terms of performance but lose touch with deeper meanings of their own behavior because they are preoccupied by very specific and narrow goals to be achieved. The individual subjectivity of this group of students is far from an autonomous subject because they are ultimately effective but compliant to the authorities and goals externally set for them. For those students who do not subscribe to the ideology, they are labeled as problems or troubled students. This group of students is subject to school or community service and counseling. Their individual subjectivity can develop resistance and an autonomous interior to an extent which the professional model of diagnosing individuals through standardized psychological tests replaces genuine love and organic human connection. Although there can be individual variations in how counselors interact with students, the broader context of school bureaucracy and the bureaucracies of the Ministry of Education and its branch organizations is not very conducive to the development of organic and meaningful human connections. Ironically, this limit and inadequacy can serve as an opening for producing individual subjectivity not totally engulfed by the new disciplinary paradigm. It is worth noting that those students who survived the Sewŏlho disaster were those who did not listen to the authorities and acted according to their own judgment.

References

Althusser, L. (1971). *Lenin and philosophy and other essays* (B. Brewster, Trans.). New York: Monthly Review Press.

Association of Parents for Genuine Education (참교육학부모회). (2011). *Students' human rights seen through the reformed school rules* (고등학교 개정학칙으로 본 학생인권), July 25. http://hakbumo.or.kr/bbs/view.php?id=ul_in&page=3&sn1=&divpage=1&sn=off&ss=on&sc=on&select_arrange=headnum&des=asc&no=400 (Accessed on September 10, 2015).

Ch'ŏn, S.-Y., et al. (2013). A study of how to make the subject-based classroom system effective: a focus on administering the curricula, class, and the guidance for students' lives in the general high school (교과교실제 운영 효율화 방안: 일반계 고등학교의 교육과정 운영 및 수업, 생활지도를 중심으로). CR 2013-38. A study sponsored by the 2013 Educational Policy Network. Seoul: Korean Educational Development Institute.

Chŏng, Y.-J., & Pak, H.-J. (2011). Students' human rights still far way, the school rules that suppress autonomy (아직도 먼 학생인권 (하) 자율억압하는 학칙). *Kyŏnghyang Newspaper* (경향신문), March 7.

Cookson, Jr., Peter, W. & Caroline H. Persell. (1985). English and American residential secondary schools: a comparative study of the reproduction of social elites. Comparative Education Review *29*(3): 283–298.

Daejeon Metropolitan Office of Education. (2011). *A consultation report on 2012 high-school curricula: a pursuit of school diversification and autonomy for managing school curricular* (2012 년 고등학교 교육과정 컨설팅 보고서: 학교의 다양화, 교육과정 운영의 자율성 추구). Daejeon City, South Korea: Office of Education.

Deresiewicz, W. (2014). *Excellent Sheep: The miseducation of the American elite and the way to a meaningful life.* New York: Free Press.

Deslandes, R. Paul (2005). Oxbridge Men: British Masculinity and the Undergraduate Experience, 1850–1920. Bloomington and Indianapolis: Indiana University Press.

Elias, N. (2000). *The civilizing process: Sociogenetic and psychogenetic investigations* (revised ed.). Oxford: Blackwell.

Foucault, M. (1979). *Discipline and punish: The Birth of the prison.* New York: Pantheon Books.

Han, T.-D. (2000). School discipline and schooling effect (학교훈육과 학교효과). *Journal of Educational Research (교육연구), 10,* 43–58.

Han, S.-J. (2014). Students of Tanwŏn High School, feeling guilty about even laughing (단원고 학생들, 웃는 것에 죄책감 느껴). SBS, April 28. http://www.mediagaon.or.kr/jsp/sch/common/popup/newsviewpopup.jsp?newsId=08100301.20140428100000113&print=Y&highlight= (Accessed on November 17, 2015).

Hong, W.-P., et al. (2013). A study of validity and flexibility of the curricular of the general high school to realize education for happiness (행복교육 실현을 위한 일반고등학교 교육과정 적합성 및 유연성 제고 방안). Publication registration no. 11-1342000-000037-01. Seoul: The Ministry of Education.

Hwang, C.-S., et al. (2013). An analysis of the outcome of the high school diversification policy and a plan to improve it (고교다양화 정책의 성과분석 및 개선방안 연구). Research Report 2013-05. Seoul: Korea Educational Development Institute.

Kalton, G. (1966). The Public Schools: A Factual Survey. London: Longman.

Ko, Y.-J. (2014). The difference in student's self-determined motivation for study related to their trust in teachers based on their perception of teachers' disciplinary types (고등학생이 지각한 사훈육유형에 따른 교사신뢰와 자기결정성 학습동기의 차이). Master's Thesis. Graduate School of Education, Pusan National University.

Korean Federation of Teachers' Associations (KFTA). (2012). Report material in the KFTA News (교총소식), "The Federation, firm refusal to the forceful implementation of Seoul students' human rights ordinance, beginning the movement to oppose the school rules reform" (교총, 서울학생인권조례 강행 단호히 거부, 학칙개정 반대운동 시작), January 31, 2012. http://kfta.or.kr/news/view.asp?bName=news&page=64&search=&search_field=&search_value=&s_div=1&num=4750 (Accessed on September 07, 2015).

Ku, C.-Ŏ., et al. (2008). *High school education in the world* (세계의 고등학교 교육). Korea Education Forum World Education Series 1. Seoul: Sinjŏng.

Kwak, N.-H. (2010). The prohibition of corporeal punishment being settled better than expected (곽노현, '체벌금지 예상보다 잘 안착'), November 16. http://news.sbs.co.kr/news/endPagePrintPopup.do?news_id=N1000815321 (Accessed on September 07, 2015).

Kyŏnggi Provincial Office of Education. (2015). *2015 Basic plan for live human rights education: making of a students-centered educational community culture* (2015 생활인권교육 기본계획: 학생중심의 교육공동체문화 형성). Enhancement of education material 2015-2 (장학자료 2015-2호). Suwon City: Kyŏnggi Provincial Office of Education.

Kyŏnghyang Newspaper. (2010). The entire text of the proposal for Kyŏnggi Province students' human rights ordinance (경기도 학생인권조례안 전문)'. http://news.khan.co.kr/kh_news/art_print.html?artid=201009171559471 (Accessed on September 10, 2015).

Moon, S. (1997). Begetting the Nation: The Androcentric Discourse of National History and Tradition in South Korea. In E. Kim & C. Choi (Eds.), *Dangerous women: Gender and Korean Nationalism* (pp. 33–66). New York: Routledge.

Moon, S. (2005). *Militarized modernity and gendered citizenship in South Korea*. Durham: Duke University Press.

Mun, S.-J. (2015). A year after Sewŏlho disaster healing beyond conflict—Interview with Un-ji Kim, school doctor at Tanwŏn High School: day by day, kids are doing their best (세월호 1년 갈등을 넘어 치유로—인터뷰 단원고 스쿨닥터 김은지씨: 아이들 하루하루 최선을 다해요). *Kungmin Daily* (국민일보). April 15.

National Labor Union of Educational Employees (NLUEE). (2012a). A plaza of school reform: Hungdok High School (학교혁신 한마당: 흥덕 고등학교). In *Proceedings of the 11th National Conference for Practicing Genuine Education*

(January 11–13, 2012). Seoul: National Labor Union of Educational Employees.

National Labor Union of Educational Employees (NLUEE). (2012b). A plaza of school reform: Iu High School (학교혁신 한마당: 이우 고등학교). In *Proceedings of the 11th National Conference for Practicing Genuine Education* (January 11–13, 2012). Seoul: National Labor Union of Educational Employees.

Pak, C.-K. (2006). Autonomous regulation of school life through students' participation (자율적인 학교생활 규정은 학생의 참여로), May 13. http://anseongnews.com/bbs/print.asp?group_name=316&section=1&category=10&idx_num=6799 (Accessed on September 08, 2015).

Ramirez, F. O., & Boli, J. (1987). The political construction of Mass schoolings: European origins and worldwide institutionalization. *Sociology of Education, 60:1*(January), 2–17.

Seoul Metropolitan Office of Education. (2014). *A comprehensive study of the making and administering of the high-school curricular in the special city of Seoul (서울특별시 고등학교 교육과정 편성, 운영지침 총론)*. Seoul: Metropolitan Office of Education.

Sim, H., Yi, P.-H., & Sŏ, T.-G. (2015). An analysis of factors affecting school maladjustment among high-school students in order to administer the academic suspension reconsideration system: Focusing on the relationships with teachers and school rules (학업중단숙려제 운영을 위한 고등학생 학교부적응 행동요인 분석: 교사 및 교칙관계를 중심으로). *Educational Culture Studies (교육문화연구), 21*(22), 57–78.

Song, C.-H. (2014a). It is O.K. to say that it is not O.K. (괜찮지 않다고 말해도 괜찮아). *SisaIN* (시사인). April 28. http://www.mediagaon.or.kr/jsp/sch/common/popup/newsviewpopup.jsp?newsId=06100252.20140428100000039&print=Y&highlight= (Accessed on November 17, 2015).

Song, C.-H. (2014b). Can we return to the time before Sewŏlho disaster? (세월호 이전으로 돌아갈 수 있을까?). *SisaIN* (시사인). July 4. http://www.mediagaon.or.kr/jsp/sch/common/popup/newsviewpopup.jsp?newsId=06100252.20140704100000002&print=Y&highlight= (Accessed on November 17, 2015).

South Kyŏngsang Provincial Office of Education. (2013). 'The standard program for various regulations to guide students' lives (학생 생활지도 제규정 표준안)'. 경남교육 2013-099. Ch'angwŏn City: South Kyŏngsang Provincial Office of Education.

Yi, C.-Y. (2003). Guidance for students' lives in primary, junior high, and high schools, 'students'-centered' change (초중고 학생생활지도 '학생중심' 전환), *Korea Economy* (한국경제), February 12. http://www.hankyung.com/news/app/newsview.php?type=2&aid=2003021275368&nid=910&sid=010620 (Accessed on September 08, 2015).

Yi, T.-Y. (2010). Remember me, revelation for children who were excluded from Sewolho (리멤버 미: 세월호에서 배재된 아이들을 위한 묵시론). *Yonhap News* (연합뉴스).

Yi, C.-H. (2014). Ansan Hospital at Korea University, surviving student of Tanwŏn High School, again aggravation of the post-traumatic stress disorder (고대안산병원, 단원고 생존학생, 외상후스트레스 재악화). *Money Today* (머니투데이). December 9. http://news.mt.co.kr/newsPrint.html?no=2014120909325646847&type=1&gubn=undefined (Accessed on November 17, 2015).

Author Biography

Seungsook Moon is a Professor of Sociology at Vassar College where she chaired the Department of Sociology and directed the Asian Studies Program. She is the author of *Militarized Modernity and GenderedCitizenshipin South Korea* (Duke University Press, 2005; reprint 2007) and a coeditor of *Over There: Living with the U.S. Military Empire from World War Two to the Present* (Duke University Press, 2010), to which she also contributed Introduction, Conclusion, and three single-authored chapters. She is a recipient of the Fulbright Scholar Award and the inaugural Sang-kee Kim Visiting Professorship at Harvard University.

CHAPTER 8

Foreign to Disaster or New Point of Solidarity? A Vietnamese Victim's Family in the Sewol Aftermath

Hyunok Lee

Who belongs in a national community? Who are legitimate members of a country entitled to the rights and safety provided by the state? How is the boundary constituted that includes some as citizens and excludes others? And how permeable is it to social solidarity that emerges in response to state failures to provide or protect? Focusing on a Vietnamese "multicultural family" who were aboard the Sewol,[1] I address these questions in this chapter as I analyze the experiences of Vietnamese family members in terms of multiple boundaries of political and economic belonging. They are transnational family members whose lives are managed beyond a national boundary. They are tied to Koreans by kinship, but are not "Korean" in terms of nationality, ethnicity, and

[1] While the majority of the victims were students from Danwon High School, there were a number of passengers who were not high-school students.

H. Lee (✉)
Yonsei University, Wonjoo, South Korea

J.-J. Suh and M. Kim (eds.), *Challenges of Modernization and Governance in South Korea*, DOI 10.1007/978-981-10-4023-8_8

citizenship. The Sewol ferry disaster not only took their family members away, but also destroyed the economic basis of their global survival strategy. On the basis of the interview I conducted with a marriage migrant's parent and sister in Ansan in 2015 at the shelter where they stay, together with my observation during the participation of the various events on the Sewol ferry disaster and an examination of the public discourse on the bereaved families, I shed light on the tension between the highly individualized yet transnational mode of a family's economic survival and national politics of belonging (and its dissolution) in a post-disaster situation.[2]

First, I briefly discuss existing studies on the Sewol disaster to highlight the necessity to examine the relationship between the state and citizenship as revealed in the incident. Second, I recount the experiences lived by the Vietnamese family members since they came to Korea after their daughter and her family had sunk with the ferry. Third, I explore how they formed their global household as a survival strategy based on the interview. Then, I discuss the disruption of their global household and the rights of the transnational family. Finally, I raise a question of social solidarity in an individualized society by exploring the politics of compensation.

The State and Citizenship in Disaster

Of a wide variety of analyses on the causes of the Sewol's sinking, a majority focuses on physical problems of the ferry and endemic corruption such as the decrepit ferry, the illegitimate reconstruction of the ferry, the overloaded freight, the lack of ballast water, and the loose regulation of the safety examination system (Ji 2014). These may be the causes of the ferry's sinking, but other factors are called for to explain why passengers were not rescued before the ferry sank completely. Other studies thus suggest that failures of human behavior such as the absence of a control tower, ineffective rescue activities of the Coast Guard, the outsourcing of the rescue activity to a private company, and

[2]A Vietnamese translator translated the interview from Vietnamese to English. Prior to this interview, I participated in the human rights committee's official interview with this family as an observer, participated in multiple events, and had casual conversation with Vietnamese family members, Korean volunteers, and Vietnamese volunteers in Kwanghawmoon and Ansan in 2015. I also interviewed a Korean volunteer, who is assisting this family while they are in Korea.

the irresponsibility of the captain and crew coalesced to turn the accident into a man-made disaster.

While various individuals acted in particular ways that exacerbated the accident, they did so under circumstances that had resulted from the sediments of previous choices and behaviors. And their behaviors were shaped by these structural circumstances that had their own logic. This logic is named differently by different scholars, depending on the focus of their analysis. You and Park (Chap. 10 in this volume), for instance, call it a regulatory capture that grew out of the earlier state corporatism of the Park Chong-Hee era. Lee (Chap. 2 in this volume) and Ji (2014) note that the Sewol ferry disaster revealed the contradictions of neoliberal governance that promoted the efficiency of small government and deregulation at the cost of the people's safety. The disaster proved the state's responsibility to protect its citizens to be a mere façade, raising fundamental questions about the relationship between citizens and states in regards to one's entitlement to basic safety and the meaning of citizenship.

Many scholars have in fact shown that disasters and their aftermath reveal various borders within a community that has suffered them.[3] They raise a set of questions fundamental to the community as to its boundary, membership criteria, and sources of its solidarity. Researchers have taken post-disaster situations as a site that exposes various distinctions within a community and an emergent mode of social solidarity. It is not just that victims experience a disaster differently, and their injuries or recoveries vary in their degree and nature. Their experiences are organized according to their socioeconomic positions, thus betraying fault lines in their societies that might have been concealed until the disaster. Smith (2006) points out that one's likelihood to survive a disaster depends on her race, ethnicity and social class, while others show that one's survival after Hurricane Katrina cannot be adequately understood without paying due attention to her sociodemographic characteristics (Cutter 2006; Bytheway 2007). Cutter (2006) adds that victims'

---

[3] Social implications of natural disasters has been an important topic of discussion, and it is based on the influence of human activities in causing natural disasters and the existing social distinctions or inequality which disasters reveal. Cherpitel (2001) noted that "nature's contribution to 'natural' disasters is simply to expose the effects of deeper, structural causes—from global warming and unplanned urbanization to trade liberalization and political marginalization" (cited in Jackson 2006).

structural vulnerability, in addition to the failure of the US government's emergency management, needs to be taken into consideration when we discuss the inadequate responses to the hurricane's aftermaths. In Japan, the great earthquake of March 11, 2011 was followed by old and new politics on citizenship, as Shindo (2014) points out. While an imagined community as "one unified entity surrounded by a boundary which separated the safe inside from the dangerous outside" gained currency in a post-disaster Japan, a new form of inclusive community emerged as immigrants participated in post-disaster relief activities. The presence of immigrants unsettled the old politics of exclusion associated with the nation-state while at the same time gesturing at a new politics of inclusion.

In a highly individualized society that many have closely associated with modernization (Beck 1992; Elliott 2002), what holds a community together? Elliott and Turner (2015) offer Solidarity and solidarity as a clue. While they associate Solidarity with the national boundary and the institutional mechanism to realize social cohesion within the Keynesian welfare state such as citizenship, they relate solidarity to shared moral values at various scales. Solidarity has been challenged in recent years. As the Keynesian welfare state has declined and globalization has progressed, so the tight link between participation in the national economy and the social and political rights of a member of the national community, i.e., citizenship, has weakened. Not only does disaster lay bare the workings of Solidarity and solidarity in an affected community that Elliott and Turner (2015) discuss, but also complicates them by bringing to the fore latent lines of division and new sources of cohesion.

Strangers in Paengmok Port

To state that the sinking of the Sewol ferry took 295 lives and nine are still missing as of June 2016 is to erase many complications generated by the disaster. The incident divided two families, one between the dead, the living, and the missing and the other between the dead and the living, and brought together Vietnamese and Koreans in familiar yet complicated encounters. Phan T. (age 31), a Vietnamese woman who married a Korean man in 2006, was on the ferry with her husband and their children. She was found dead on April 23, 2014. Her husband (52) and son (age six) are stilling missing, and her daughter (age five) is the only survivor. Her father, Phan V. (age 63), and younger sister, Phan H.

(age 26), were living in Vietnam at the time of the incident, and came to Korea after they received a call the Korean Consulate made to inform them of the accident.[4] Upon their arrival in Korea in April 2014, they tried to understand how the accident happened and why Phan T. had to die, but soon learned it would be a challenging process, not least because of the language barrier.

They were initially provided with a translator by the government[5] until Phan T.'s dead body was retrieved, but have been on their own most of the time since then. Even though they occasionally received help from other Vietnamese marriage migrants or Vietnamese students in Korea who volunteered to translate for important meetings or interviews, they, with the volunteers' occasional assistance, were not able to understand the situation or follow new developments. "I was so frustrated with the communication problem. Nobody explained to us what was going on. I started to teach myself Korean," Phan H. (25) said in Korean. They listed communication as the biggest difficulty they experienced since they came to Korea. They were often excluded from latest updates for and decision-making by the Sewol bereaved families. When the bereaved families agreed on terminating the Coast Guard's operation to search for the missing people in November 2014, for example, the Vietnamese families were not fully informed, much less consulted. It was only after reading Vietnamese news that they learned of the agreement. "We are also a bereaved family. Why does nobody ask our opinion?" asked Phan H. An immediate answer had to do with the language difference that encumbered the communication between them and the other bereaved families.

There were, however, deeper and more complex problems than the language barrier. They were a bereaved family through kinship, and *foreigners* in Korea. Phan T.'s brother-in-law, a Korean who worked with other bereaved families to share information and make decisions, admitted that he was not very good at sharing information with Phan's family (Heo 2014). If he had been "good" at sharing information with the Vietnamese family and brought their voice to the decision-making process as a representative, Phan's family may have felt less marginalized.

[4] I identify the interviewees only with their initials to protect their privacy.

[5] Interviewees mentioned that the translator was provided by the government but they were not able to name which government agency it was.

And yet he did not, and whether it was because he either was unable or refused to do it, his inaction deprived the Vietnamese family of access to information and left little room for them to raise their voices. He and they were given unequal access to relevant information: he enjoyed the same access that the other Korean families did whereas they were denied this. He was empowered to participate in the same decision-making that the other Korean families were whereas they were not. In a way, his empowered position reflects the power relationship within the family which is the combined results of patriarchal relationships and Phan's family's foreign status. All the bereaved families suffered a tragic loss of their beloved ones, and yet they were differently treated according to their language and citizenship. The post-Sewol experiences laid bare a social hierarchy, structurally associated with the difference in citizenship, as the Vietnamese were forced to climb a steeper hill to gain access to information and decision-making.

It is granted that communication was a big problem in Paengmok Port, the most serious of which existed between the government and the bereaved families. Because the government was not forthcoming with information about the accident or the subsequent rescue operation and because when it offered information, it was not credible, most of the victims' families had difficulties in understanding the situation. It can be safely said that none of the family members in Paengmok Port found adequate the level and scope of information provided to them. And yet it may be as safely said that the post-disaster situation the Vietnamese family encountered was much harsher. At least one of the sources of their extra difficulties may be traced to their citizenship.

Transnational Family and Global Householding

Why was Phan T., a Vietnamese by birth, in Korea in the first place? The seemingly simple question leads us to inquire into larger forces at work that have transformed the countries of her birth and living (and death) and beyond. She had come to Korea in 2006 to marry a Korean man, as many Vietnamese women did in the 2000s. Although there have been marriages between Koreans and non-Koreans before, a particular kind of international marriage began in the late 1980s to find brides for bachelors in rural areas. As a result of Korea's rapid industrialization and urbanization, many young people, particularly females, had migrated to urban centers in search of employment in the 1960s and 1970s. Most of those who were left in the countryside were male, old, undereducated, low-income, and thus unattractive as a mate. By the 1980s, many were singles who had few prospects to find a Korean bride, and thus turned their eyes overseas (Lee 2012). This rural phenomenon grew common in the urban area during the 2000s, as 58.6% of multicultural families reside in Seoul Metropolitan area as of 2012 (Jeon et al. 2013). The husbands of this international union, whether in rural or urban areas, have on average a relatively low level of educational attainment, a low income, and an older age in terms of marriage (Jones and Shen 2008). Korea's development path, combined with the lack of public care provision that would facilitate their social reproduction, in other words, had generated a group of doubly disenfranchised bachelors who had to devise a survival strategy on their own (Lee 2012).

This group of bachelors in Korea was able to find a means to social reproduction overseas thanks to changes outside their border. China and then Vietnam followed Korea as late developers to develop their economies in a globalizing world, and their modernization too was accompanied by a massive migration of their citizens, who left countryside in search of better economic opportunities. Just as some Koreans left their country to find employment in the United States or Germany in the 1970s and just as Filipinos left theirs for better opportunities in the USA or Japan in the 1980s, Chinese and Vietnamese started their international migration in the later decades. Many Chinese women left their home in the countryside to marry a Korean man in the 1990s, and many Vietnamese women followed in their footsteps the following decade.[6] The influx of Vietnamese brides became noticeable during the 2000s and they soon became the second largest group after Chinese. By leaving home, they helped themselves as well as their families left behind. Not only did they earn a living on their own in a foreign country but also remitted part of their income to their families. International marriage appears to be a household strategy for Vietnamese families who had been marginalized from the recent economic growth of their home country and left on their own after welfare reform (Lee 2013, 2012).

[6] See Lee (forthcoming) for more explanation.

The confluence of socioeconomic conditions in Korea and Vietnam, in short, brought Phan T. and her husband, Kwon, and many other transnational couples, together in union. Their international marriage seemed a collective means to social reproduction for low income households in Korea and Vietnam where public care provision was limited (Lee 2012). At the same time, the formation of "multicultural families" in Korea reflected the regional economic disparity where the husband's relatively lower status in Korea represented a relatively better opportunity for the Vietnamese woman.

The phenomenon of international marriage is not unique to Korea, as many societies are gradually incorporated into a globalizing world, as many scholars have shown. Emphasizing the growing importance of a household in the global political economy, Safri and Graham (2010) define the "global household" as "an institution formed by family networks dispersed across national boundaries" characterized by "ties of economic and emotional interdependence and structures of decision-making or governance" despite the geographical dispersion of their members (ibid: 100, 107). Douglass (2006) refers to "global householding" as "a continuous process of social reproduction that covers life cycle stages and extends beyond the family" (ibid: 423). In some sense, global householding is a way that people with fewer resources cope with their current economic problems in countries where individuals assume increased responsibilities for their survival. A "multicultural family" in Korea, in this sense, represents a union of two global householdings, a Korean one and a Vietnamese one, negotiating the global political economy together. And yet, the union is not made on an equal footing, a global, or at least regional, hierarchy as Lee et al. (2016) noted.

My interview with Phan T.'s family shows how they formed a transnational family and engaged in global householding. Phan V. was a fisherman in the southern part of Vietnam, and his family was never affluent. Phan T, the smartest among Phan V.'s five children, moved to Ho Chi Minh City to find a job. While she worked at a garment factory, she learned about an opportunity to marry a foreigner.[7] She participated in a meeting, organized by a commercial marriage broker and met the

[7] It was a commercially arranged marriage brokerage, often called a "bride contest". It was subject to criticism, as it was associated with trafficking in women. See Lee (2014) for more details.

Korean man whom she would eventually marry.[8] She moved to Korea in 2006 when she was 23 years old. For the subsequent 10 years, she and her husband supported her family in Vietnam. It certainly helped her Vietnamese family to improve their livelihood. However, when Phan V. visited Korea for the first time in 2011, he realized that the support that his family in Vietnam received was "the sweat of his daughter and son-in-law". His son-in-law worked at a construction site and his daughter was a full-time housewife taking care of two children at that time. He suggested his daughter work outside home. Since then, both his daughter and son-in-law managed a cleaning business together. Like many other multicultural families, they sent remittances to their family of origin and marriage migrants often participate in the labor market to support families both in Korea and their country of origin (Lee 2013).

The Phan family was geographically spread across national boundaries, and yet their household was linked economically and emotionally. They have maintained their ties through regular phone calls and occasional visits to each other; and remittance was an important part of nurturing their relationship. While they thus constituted a global household that had to negotiate multiple institutional and cultural boundaries, they could not rely on their respective governments—much less on global governance to the extent that one exists—for public care provision. Just as Phan T.'s migration and her husband's international marriage were their individual coping strategy in a globalizing world that shifts to individuals more and more responsibilities for survival, their global household too was on its own for survival, a painful reality laid bare by the unexpected loss of Phan T. and her husband. When she passed away and her husband was missing in the Sewol disaster, their loss disrupted the Phan family's global household, exposing the vulnerability of the global household as an economic and emotional unit. Phan V. and Phan H., the Vietnamese father and sister, literally fell between the two states: they were strangers who had no legal standing in the Korean system while Phan T.'s death and their life in Korea were of no concern to the Vietnamese government. Their global householding geographically separated the household members across multiple boundaries, and yet united them not only emotionally or economically but also in their disenfranchisement from their government's protection and provision.

[8] See Lee et al. (2016) for more details.

Disruption of Global Householding and a Question of Belonging

> My daughter is a Korean citizen as she married a Korean man. I am her father. I have a right to know what happened to her. (Phan V.)

Phan V.'s statement abstracts the complex nature of their positionality in this situation. They are a naturalized citizen's family members and foreigners. After his daughter's body was retrieved from the sea, they waited for the news on their son-in-law and grandchild. The only survivor was their granddaughter whose temporary custody was taken by her aunt from the father's side. In terms of compensation, this family's situation is quite unique compared to other bereaved families as they have a survivor, two missing persons, and a deceased person within one family. While the amount of compensation may vary depending on whether their family members are missing, deceased, or surviving, the compensation for deceased family members will be by law disbursed to the five-year-old granddaughter, who is the only survivor. Whether Phan V. and Phan H. have maintained the global householding that the family in Vietnam was dependent on the couple in Korea for survival has virtually nothing to do with the question of who will be compensated for the loss of their loved ones and the source of income. When I consulted this with legal advisors, they quickly jumped into the question of who was taking care of the surviving granddaughter. In this case, they had a right as linear ascendants, but it was not clear how the fact that they were foreigners with fewer resources might play out in the court. Important as it was to decide on how much the compensation should be and who was entitled to it, these questions grew out of proportion and consumed much of the post-disaster public discourse. How was it then that the public discourse was reduced to a matter of monetary compensation? More directly relevant to the chapter's topic, how was it that the issue of compensation quickly became grounds for criticism and exclusion of this family?

Politics of Compensation and the Ultimate Individualization of Survival

> Some people ask us why we stay in Korea for so long. Some people say that we are so greedy. Some people ask us what we are going to do with the (compensation) money. (Phan V.)

During their stay in Korea, Phan V. and his youngest daughter were often confronted with a challenge that they would not have been if they had not been foreigners. Their motive was put on a public trial. They were accused of staying to get the compensation or to settle in Korea permanently. As they suffered from these accusations, they reacted quite sensitively to the question of why they stayed in Korea and what they wanted to see out of the tragedy eventually. They answered that they wanted to see the funeral of the three people and a secure environment for the granddaughter to grow up. Like other bereaved families who cannot leave Kwangwhamoon, a site of their ongoing campaign to unearth the causes of the Sewol disaster, they could not just go home and live their lives as if nothing happened unless they brought closure to their loss somehow.

Compensation is an important part of the post-disaster discussion that raises complex questions: "Is it appropriate to use financial compensation to repair harm? How should pay and on what basis should the compensation be awarded? What is the nature of the damage? How to evaluate it and how to value the amount of compensation determined?" (Barbot and Dodier 2015: 81). Moral tensions around the process of compensation deserve more attention as compensation occupies an important position in the discussion of a citizen's right to safety. Exploring victims' reaction to the compensation process for the iatrogenic human growth hormone, Bardot and Dodier (2015) outlined three dimensions of moral tensions around the process of compensation: (1) expectation of justice; (2) expectation of suitable compensation; and (3) tensions reactivated by the attitude of a third party. Bardot and Dodier (2015) offered more explanations on each of the dimensions as follows. First, victims were generally suspicious about the government's attempt to provide for compensation. Some families perceived "the compensation as part of a possible strategy to silence them, to distract them from their search for the truth about the circumstances of the tragedy, and to avoid those responsibility from being brought justice." (Bardot and Dodier 2015: 86). Second, was a question on the incommensurability of the damages and compensation after the tragedy. As the compensation was associated with their children's death, it was very difficult for the families to accept it. This was a different moral tension from the first one but they often emerged in a combined form, "avoid being bought off" and "avoid receiving the price of child's life" (Bardot and Dodier 2015: 88). Third, there was a tension created by third parties as the case was publicized.

This exposure to public opinion generated unwarranted judgment from which victims suffered.

The issue of the Sewol victims' compensation was not immune from the three moral tensions discussed by Bardot and Dodier or from national politics. First, a compensation plan was announced by the government when the bereaved families and the general public were demanding a thorough investigation of the disaster to uncover the causes of the disaster. If a family accepts compensation, that would mean reconciliation between the state and the family who accepts it, according to the special law on Sewol Ferry (Article 16). It means that once bereaved families accept the compensation, they cannot raise issues any more. The moral tension stemming from the expectation of justice surfaced as a division between the families who decided to accept compensation and those who refused. As of September 30, 2015, the last day of submitting the application for the compensation, 68% of the bereaved families submitted the application. Those who chose not to submit the application demanded justice and filed a lawsuit against the government and Chonghaejin, the owner of the Sewol, for damage (Jung 2015).

Second, how the amount of compensation was calculated and who pays for this was another point of moral tension. According to the Ministry of Oceans and Fisheries, a victim of the Sewol disaster was expected to receive more than US $820,000[9]: $420,000 for disaster compensation from the government, $300,000 from funds raised by the public, and $100,000 from private travel insurance. The amount of compensation by the government was calculated by estimating how much a victim, had he not died in the accident, would have made for the rest of his life under the assumption that his income would be similar to that of a day laborer. The assumption was arbitrary and arbitrarily low, conflicting with the expectation that at least some families had of their children. Some were outraged that their children were valued so low by the government that the children were desecrated after having been killed by its negligence. Others were troubled that they had to put a value, any value, to the loss of their loved ones.

The second tension was intertwined with the attitude of a third party in this case. There were growing complaints about the management of

[9] Count US $1 as 1000 Won.

the post-Sewol disaster that compensating the bereaved families was like abusing tax. It is notable that the public's attitude towards the Sewol families changed from mourning and sympathy to annoyance and concerns about favoritism after the government brought up the compensation issue (Cheon 2015). Cheon (2015) observes an increase in public concerns that compensation would amount to preferential treatment of the bereaved families, and that yielding to the families' demands would add an unfair and illegitimate financial burden to the government—and ultimately to the taxpayers.[10]

These concerns by third parties were grounded at least partially in the first two moral tensions. As soon as the issue of money was raised, the families' sincerity was questioned, and their grief seen with suspicion. Bereaved families were accused of getting easy money in exchange for their family member's death. Now the tables were turned. The families became a main target of criticism. Granted that the compensation process was riddled with inherent moral tensions, third-party interventions—the ones particularly by right-wing groups such as Oeboei Yeonhap and Ilbe—played a direct role in channeling the nature of public discourse in a particular direction: away from sympathy toward skepticism. The rise of the third parties' hostile words that portrayed the bereaved families as selfish actors who sought to maximize their gains narrowed the little room there was for public discussion on citizens' right to safety and the state's responsibility.

Thus, the politics of compensation can be considered in three strands. The first moral tension resulted in the division among the bereaved families between those who took compensation and those who refused to take it. The second one weakened the moral standing and negotiating position of the first group, while the third one shifted the moral burden from the Korean state to the families. The moral high ground that the third parties took over by highlighting the first two moral tensions limited the political imagination for social justice and solidarity by appropriating the language of fairness and reframing the post-disaster discourse in monetary terms.

[10]It is worth noting that these were unwarranted concerns because the compensation was entirely legal and even if the government paid the compensation first, it would later collect the amount, together with the fees, from Chonghaejin, as it did in November 2015 (Kim 2015).

Rights of Transnational Families in Disaster

The accusations that the Phan family intends to stay in Korea for compensation or permanent settlement may be partly understood in the context of these moral tensions and politics around the compensation. Also, they raise a fundamental question about what holds a community together. Before the issue of a citizen's right to safety can be brought up for public discussion, there must exist shared ideas about social justice and solidarity. The fact that there were tensions, divisions, and diversions surrounding the compensation should not belie the reality that a bereaved family could, if they so chose, make a claim for compensation, a reality grounded in shared ideas about social justice and solidarity. The idea is accepted as so commonsensical within a given society that it remains invisible under normal circumstances. If one's claim to rights is open to question, as the Phans' was in Korea, however, their status brings to light the boundary embedded in the shared ideas. Although political economic changes in both Korea and Vietnam steered individuals on the lower rung of their respective societies to form a transnational family as a household reproductive strategy, the survival strategy ran the risk of falling between the two states in the aftermath of a disaster, for both could deny responsibility for protecting the individual. Individuals had to go beyond national boundaries to design a survival strategy in individualizing globalization, and yet public service provision remained organized within national boundaries. The Phans' experiences accentuate the unevenness in globalization that doubly disenfranchises those on the margins: once by pushing them out of their national boundary for survival and once again by leaving them trapped in between national boundaries when they need public assistance.

Finally, the Phans' experiences point to another complication in the issue of solidarity. The moral tensions and some people's hostility towards the Sewol families notwithstanding, more than 800 civil organizations across the country participated in the People's Council for Demanding Justice for Sewol Disaster and various forms of groups continue to engage in public discussion on human rights and safety at the grass-roots level. Despite many challenges confronting them, the Phans were able to stay in Korea and continue to raise their voice as a bereaved family thanks to assistance and solidarity from civil society.

> People think that we have a large amount of money for compensation from the Korean government and stay in Korea comfortably, but that's not true. The Korean government didn't do anything for us. We were able to stay in Korea because of the support of good Korean people.

Phan H. and Phan V. emphasized that they were able to stay in Korea thanks to "good Korean people" who supported them, not the Korean government. A more accurate description would be that the Korean government provided basic support, and this was obtained thanks to the demands and hard work of good-willed Korean people. The Korean Consulate in Ho Chi Minh City gave a call to inform them that Phan T. was on the ferry, and Ansan City government assisted them through the Center for Multicultural Families. But the Community Chest of Korea raised the funds to pay for their airplane tickets, and the Writers' Association assigned two writers to assist them. Many individuals volunteered to assist them to secure their visa status and raised their voices as a part of organized bereaved families. Despite the many difficulties they faced in Korea, it is remarkable that they were able to stay in Korea at all and participate in many activities as a part of the bereaved families for nearly 2 years. It would have not been possible if there had been no organized solidarity by civil society. It is probably too optimistic to consider this a new emergent form of inclusiveness in Korea, and yet it deserves attention as a potential basis for a new type of solidarity in Korean society.

Conclusion

The aftermath of the Sewol disaster raised fundamental questions on citizens' rights to safety, the state's responsibility, and social solidarity in South Korean society. Not only did the disaster show that Solidarity, the institutional mechanism for guaranteeing citizens' rights to safety as a part of social cohesion, rarely worked, but also the various moral tensions around compensation raised challenges on the idea of shared moral values. In particular, the experiences of a Vietnamese family caught in the disaster revealed the complexity of this question in the global context by demonstrating that one's survival strategy been both individualized and globalized in recent decades, yet they were left in the cracks between the institutional mechanisms of two countries when their global household was disrupted. In other words, there is little room for them to claim their rights in this kind of situation. They made efforts to improve their marginalized economic condition by using the strategy of global householding, yet the disruption of the basis of this global household revealed the precarious nature of this household. In a way, the Phan family was marginalized economically as well as politically.

Although the aftermath of the Sewol disaster, in particular the politics of compensation, raised skepticism on social solidarity in South Korea, social solidarity plays an important role in explaining the experience of the Phans. Under the circumstances of little institutional support as Solidarity for them, the only source of support that they could rely on was shared moral values as solidarity. Some people who understand the Phans' unique situation and recognize their rights as a bereaved family made efforts to get institutional support from the government. This can be seen as one example of how social solidarity works. It also provides a clue for how the substance of social solidarity can be cultivated in a more inclusive manner.

In this regard, collective discussions and actions across civil society after the Sewol disaster deserve more attention. While I conducted research, I came across numerous volunteers who wanted to share the feelings of the bereaved families and support them. Some were in direct contact with the bereaved families; others organized discussions in their own community on this matter. How social solidarity forms and works is an important question to be explored in future and how the Phans' presence may enrich the discussion on the new source of cohesion and social solidarity needs to be taken into account in the future.

Websites

http://416act.net/
http://416family.org/
http://sewolho416.org/

References

Barbot, J., & Dodier, N. (2015). Victims' normative repertoire of financial compensation: The tainted hGH case. *Human Studies: A Journal for Philosophy and the Social Sciences, 38*, 81–96.

Beck, U. (1992). *Risk society: Towards a new modernity*. London: SAGE.

Bytheway, B. (2007). The evacuation of older people: The case of Hurricane Katrina. *Understanding Katrina: Perspectives from the social sciences.* New York, NY: Social Science Research Council.

Cheon, K. (2015). Geudeuleul segeum dodukeuro mandeuneun wanbyeohan bangbeob [perfect way to make them tax stealer]. *Sisain, 395.*

Cherpitel, D. (2001). Planning recovery to minimize future risk, *World Disaster Report.* Geneva, International Federation of Red Cross and Red Crescent Societies, 6–7.

Cutter, S. (2006). The geography of social vulnerability: Race, class, and catastrophe. *Understanding Katrina: Perspectives from the social sciences.* New York, NY: Social Science Research Council.

Douglass, M. (2006). Special issue: Global householding in East and Southeast Asia. *International Development Planning Review, 28,* 4.

Elliott, A. (2002). Beck's sociology of risk: A critical assessment. *Sociology, 36,* 293–315.

Elliott, A., & Turner, B. (2015). *On Society*, Seoul: Yihaksa (translated version in Korean).

Heo, J. (2014). Beteunamin sewolho yugajok panbanjjaineun malhanda [Vietnamese 'bereaved family' of Sewol ferry says], *Hangyeore*, December 26, 2014.

Jackson, S. (2006). Un/natural disasters, here and there. *Understanding Katrina: Perspectives from the social sciences.* New York, NY: Social Science Research Council.

Jeon, G., et al. (2013). *National survey on multicultural families 2012.* Seoul: Ministry of Gender Equality and Family.

Ji, J. (2014). The political sociology of sewol ferry disaster, [Sewolho chamsaui jeongchisahuihak]. *Economy and Society, 104,* 14–55.

Jones, G., & Shen, H. (2008). International marriage in East and Southeast Asia: Trends and research emphases. *Citizenship Studies, 12,* 9–25.

Jung, E. (2015). Stopappaneun meomchuji anneunda [father does not]. *Hangyeore, 21,* 1081.

Kim, Y. (2015). Beobmubu, 'sewolho chamsa' cheonghaejin haeun imjikwon seonwon deunge gusanggwon cheonggu [Ministry of justice claimed the right of compensation to Cheonghaejin ferry], *Newsis*, November 19, 2015.

Lee, H. (2012). Political economy of cross-border marriage: Economic development and social reproduction in Korea. *Feminist Economics, 18*(2), 177–200.

Lee, H. (2013). Global householding for social reproduction: Vietnamese marriage migration to South Korea. In J. Elias & S. Gunawardana (Eds.), *The political economy of household in Asia* (pp. 94–109). Basingstoke: Palgrave MacMillan.

Lee, H. (2014). Trafficking in women? Or multicultural family?: Contextual difference in commodification of intimacy. *Gender, Place and Culture, 21*(10), 1249–1265.

Lee, H. (Forthcoming). Negotiating boundaries: Women's migration throughout the life course in the multiple contexts of economic development. *Geneses.*

Lee, H. et al. (2016). Adapting to marriage markets: International marriage migration from Vietnam to South Korea. *Journal of Comparative Family Studies, XLVII* (2), 267-288

Safri, M., & Graham, J. (2010). The global household: Toward a feminist postcapitalist international political economy. *Signs, 36*(1), 99–126.

Shindo, R. (2014). Enacting citizenship in a post-disaster situation: The response to the 2011 great East Japan earthquake. *Citizenship Studies, 19*(1), 16–34.

Smith, N. (2006). *There's no such thing as a natural disaster, understanding Katrina: Perspectives from the social sciences.* New York, NY: Social Science Research Council.

Author Biography

Hyunok Lee is an Assistant Professor at the Department of Global Public Administration in Yonsei University, Korea. She received a Ph.D. at the department of Development Sociology, Cornell University in USA. Her general research interest lies on exploring the gendered political economic processes in East Asia with the focus on social reproduction. Her current research interests include gendering migration system in East Asia, citizenship in relation to developmental state and welfare regime in East Asia, Intersection of migration regime and care/welfare regime, and social economy and gender in changing care/welfare regime in East Asia.

CHAPTER 9

From Passive Citizens to Resistant Subjects: The Sewol Families Stand Up to the State

Hyeon Jung Lee

Introduction

How did the families of victims of the Sewol ferry disaster, individuals who had been "passive citizens" under the government's control, become "resistant subjects" against state power? South Korea has had dozens of disasters over the last two decades, including the Daegu Subway Fire in 2003 (192 casualties) and the Sampoong Department Store Collapse in 1995 (501 casualties). However, no victim families of other disasters had taken such full-scale, organized, and persistent action as the families of the Sewol tragedy have done.

Indeed, the bereaved families of the Sewol ferry disaster have demonstrated an unparalleled activism in the history of social movements in South Korea. Facing the government's unwillingness to investigate the cause of the accident and the rescue failure, the families became convinced that a thorough investigation into the catastrophe would not be possible without a special law. Their goals were clear, expressed in three slogans: "uncover the truths (진상규명)," "punish persons in charge

H.J. Lee (✉)
Seoul National University, Seoul, South Korea

J.-J. Suh and M. Kim (eds.), *Challenges of Modernization and Governance in South Korea*, DOI 10.1007/978-981-10-4023-8_9

(책임자 처벌)," and "establish safe society (안전사회건설)." To pursue the enactment of the Sewol Special Act (세월호특별법), the families took a wide range of actions: they collected signatures and staged overnight sit-in protests at the National Assembly Building (119 days) and at Chung-un community service center near the Presidential Blue House (76 days); they went on hunger strikes, the longest up to 43 days; they conducted lecture tours all over the country to give firsthand accounts that the mainstream media ignored; and they organized multiple national marches. In the end, spurred on by massive civil support—6,500,000 signatures were collected—the Sewol Special Act was finally passed by the National Assembly on November 9, 2014. It was seven months after the disaster.

It was, by no means, the end of the matter. In a proposal for the Act Enforcement Ordinance (*4.16*세월호참사 진상규명 및 안전사회건설 등을 위한 특별법시행령), the government tried to reduce the special committee's authority and the scope of the investigation. Furthermore, the government announced compensation plans, which highlighted a colossal amount of money to be provided to the bereaved families. Coming just two weeks before the first commemorative anniversary of their children's deaths, the families took the announcement as an insult. On April 2, 2015, 52 parents shaved their heads in protest against the government's plans to provide compensation in lieu of a thorough investigation. One grieving father said, "I want to grab and kill the rude people who chose to talk about money on the anniversary of our children's death (Fig. 9.1)."[1]

Nearly two years since the sinking of the ferry, the collective action of the families continues. They have watched salvage operations by China's state-owned company, Shanghai Co., while sleeping in tents on the top of Donggeochado, the island closest to the accident spot. They have attended hearings held by the investigation committee constituted by the Sewol Special Act. The families have also organized picketing in front of

---

[1] I would like to express my deepest gratitude to the bereaved families who were willing to share their painful and personal experiences with me. I also thank many activists, volunteers, and local residents who have helped me better understand the post-accident situations. And finally, yet most importantly, I wish to send my heartfelt condolences and consolation to the 304 victims, especially 250 students, and their bereaved families. In this chapter, the names of the interviewees are intentionally concealed for the purpose of protecting them.

Fig. 9.1 Parents of ferry victims getting their heads shaved in protest (Photo by Inki Hong, Copyright © The Hankook-Ilbo)

the Gyeonggi Provincial Office of Education to demand the preservation of the students' classrooms. Instead of cleaning up the classrooms filled with letters and gifts from all over the nation and international communities, the families argue that the place should serve as a memorial to learn from the misguided instruction to "stay still" (가만히 있으라) and to call for educational reform in South Korea.

How, then, did the victim families of the Sewol ferry disaster become such unique and empowered political actors? Why did they refuse to comply with the micro power of the state and organize collective actions to fight against it? This chapter attempts to answer these questions through the lens of the politico-cultural contexts which have influenced the families of the Sewol tragedy and helped to transform them from passive citizens to resistant subjects. My discussion focuses particularly on two key facts: first, the indignation towards and mistrust of the government and political leaders in this country, including officials, members of the National Assembly, and the President; and second, the profound feelings of guilt over the death of their own children.

Most of the data for this study comes from over a year of ethnographic fieldwork (from May 2014 to July 2015) in Ansan, the home city of the student victims and their families, and other sites related to the families' political activities, such as the Gwanghwamun Square, the National Assembly Building, and the small port of Paeng-Mok of Jindo, a far southwestern island. During the fieldwork, I observed and participated in different political activities carried out by the families and other civil supporters. I shared part of the families' everyday lives in Ansan and conversed with them in different contexts. My discussion in this chapter, most of all, aims to reveal the families' felt experiences and unarticulated feelings. Although this chapter focuses on the group's shared experience, I recognize that the experiences and suffering of individual families—even individual parents—are unique and can never be subsumed by a collective sentiment.

Deep Indignation Towards and Mistrust of the Government and Political Leaders

The families' activism cannot be explained without recognition of the Korean government's incompetence and irresponsibility after the ferry sinking. Here the word "incompetence" or "irresponsibility" does not simply refer to the government's failure to respond promptly to the accident and to save lives. Public officials, from Coast Guard officers to the President, distorted or lied about the situation and shifted blame from one party to another. Victim families were understandably enraged.

In such a grave situation, no one assumed responsibility. At the initial stage, officials in charge never clarified the exact number of passengers and the number of survivors. A vice-minister of the Ministry of Security and Public Administration kept on changing the number of survivors in his official announcements, from 161 persons on April 16 at 11 am to totals of 368, 164, 174, 175, 176, 179, and finally 174 persons on April 18 at 10 pm.[2] Indeed, a text message from the school announcing "all are saved" (전원구조) created the expectation that families could see their children, alive though frightened, when they arrived at Jindo. However, what they instead discovered was the fact that Korean officials did not even know the number of passengers in the ferry. Facing the rage of the families, the vice-minister made matters worse by ceasing official announcements of the list and shifting responsibility to other agencies.

Ultimately, what led the families to an almost primal fury was the fact that government officials did not hesitate to lie. In the nervous and anxious situation, they seemed to be less concerned about the victims and

[2] *Newsis*, April 22, 2014 "Waving numbers, growing distrust of the government" ("'파도치는 집계 숫자'에 쌓이는 정부 불신"). http://www.newsis.com/ar_detail/view.html/?ar_id=NISX20140422_0012872065&cID=10201&pID=10200.

their families, but more concerned about how they pleased their superiors. Indeed, the deception reached a peak when officials inflated the number of personnel dedicated to rescuing survivors. On April 16, the Ministry of Security and Public Administration announced that 350 special rescue crews were undertaking the task, while the families at the port could see only a few boats roaming over the sea. And the equipment which the government promised—31 helicopters, 60 ships, and 46 rescue boats—never appeared at the scene. The officially announced number of special rescue crews increased to approximately 400 on the second day and to more than 500 on the third day.[3]

On April 18, the third day of the disaster, the commissioner of the Korea Coast Guard (Maritime Police) announced that the number of "deployed" divers was as many as 500; while the announcement was broadcasted over the nation, not even one tenth was there. At the hearing on December 15, 2015, when asked about the discrepancy in the statement, the commissioner said with composure: "Deployment (투입) is different from diving (잠수). Five hundred is the number of those who were mobilized from all over the nation, while actual diving was operated by two persons at a time." He made no apology and offered no regret for the confusing numbers. Instead, he tacitly blamed the families and the public for their lack of expertise.[4]

The commissioner's statements were just a small example of government obfuscations. Other officials, whether at a low or high levels, also told similar lies without shame; their attitude revealed their condescending perception of the public and the families. The families came to realize how the state viewed them, categorically "powerless people" (힘 없는 사람들). It became obvious that the state regarded ordinary people as little more than a mob who could be easily manipulated for political purposes. The families came to understand that it is not the people and their lives that the state is primarily concerned about; the state and its officials care first and foremost about their own power.

[3] *YTN news*, April 16, 2014 at 2 pm "The fourth briefing of the Central Disaster Countermeasures Headquarters" ("여객선 사고 관련 중앙재난대책본부 발표, 네번째 브리핑"). https://youtu.be/MEq0a_xdRJI.

[4] *Pressian*, December 15, 2015 "The former commissioner of Korea Coast Guard had said 'deploying 500 divers' turned out to be a lie" ("전 해경청장, '잠수사 *500*명 투입' 거짓말 발각"). http://www.pressian.com/news/article.html?no=131835.

It was from these realizations that the political awakening among the bereaved families of the victims emerged. They regretted and reflected deeply on their past attitude toward social activism; in the past, they had distanced themselves from activism, regarding it as a job for other people. With all the government's disappointing responses, however, the families gradually found themselves changing their views and willing to live differently, based on the lessons they had learned from the post-disaster experiences: only political action will change the attitude of the state toward people; only political action will be able to stop the state from treating the people with disdain and violence.

It was, however, not only government officials who frustrated the bereaved families. Members of the National Assembly, particularly those who belonged to the ruling party, showed little willingness to listen to the families' voices; they became even less willing after the ruling party's victory in local elections on June 4. Politicians, whether of the ruling conservative or the leading progressive opposition party, betrayed the families by failing to live up to the norms of representative democracy. If there were any differences between the two, it was just a matter of "when" and "how" the betrayal happened.

Indeed, having gained confidence after the election, the ruling party immediately changed its stance toward the parliamentary investigation of the government administration and used it to its advantage. The parliamentary investigation had originally been set up to probe into the cause of the Sewol ferry accident and the rescue failure, but it was to no avail. A total of 90 days given for investigation work passed with mutual invective between the two major parties. Watching this perplexing situation in despair, the families eventually decided to stage overnight sit-in protests at the National Assembly Building. Their experience and intuition told them that the mission of the Special Act would never be fulfilled and the truth of the disaster would never be known without public pressure.

And yet the families' sit-in protests failed to change the politicians' attitudes. In political negotiations over the content of the Special Act, the two main parties either ignored or restricted the families' voices, and the effectiveness of Special Act was ultimately limited.

The key issue was whether the power to subpoena and prosecute would be given to the special investigation committee. From the family' perspective, the committee should have had the power of independent investigation and accusation to unearth the truth. However, reluctant to put themselves in a legal quagmire, the ruling party opposed granting these two powers. They argued that their inclusion in the Act would threaten the current judicial system, although 229 Korean jurists publicly declared that such powers would not infringe upon the constitution.[5] Meanwhile, the leading opposition party, which initially agreed with the families, secretly accepted the ruling party's opinion on August 7, because of the concern that its persistence on this particular issue would negatively affect its future electoral prospect. Needless to say, the sudden agreement between the two parties shocked the families, who once again felt betrayed.

In addition to the legal maneuvers, politicians hurt the families in other ways. From the very beginning, conservative politicians made misleading attacks on the families such as denouncing them as gold diggers or pro-North Korean leftists.[6] During the families' sit-in at the National Assembly Building, some politicians treated them with contempt and hatred. They shouted at the families on hunger strike not to profane the sacred place and to go away. Members of the leading opposition party, on the other hand, acted opportunistically while prioritizing their own political interests. From time to time, progressive politicians stopped by the protests and encouraged the families' resistance but did little more. For the bereaved families, these politicians did not live up to their roles as the people's representatives.

The experience of working with politicians on the Special Act gave critical momentum to the bereaved families' political awareness. The families began to see a fundamental problem in the Korean political system; the representatives expressed little interest in the people's needs and requests; politicians instead focused on the outcome of elections. Over time, the families came to realize that only direct political participation could solve the problem and fulfill the people's will. While unsatisfactory to many, the Sewol ferry Special Act was finally passed by the National Assembly on November 7. Upon this development, the families stopped their sit-in protests both at the National Assembly Building

---

[5] *Ohmynews*, July 28, 2014, "229 Korean jurists, 'the investigative power in the Sewol Special Law is constitutional'" ("법학자 *229*명 '세월호 특별법 수사권, 헌법상 문제 없어'"). http://www.ohmynews.com/NWS_Web/View/at_pg.aspx?CNTN_CD=A0002018015.

[6] *Ohmynews*, April 20, 2014 "Congressman Han Giho, "Sinking of the Sewol ferry, the left-wing groups should be weeded out" ("한기호의원 '세월호침몰, 좌파단체 색출해야"). http://www.ohmynews.com/NWS_Web/view/at_pg.aspx?CNTN_CD=A0001982806.

and at the Chung-un community service center near the Blue House, the President's residence. However, the fall of 2014 was only the beginning of a long political struggle.

Like many other powerless people in the contemporary history of Korea, the families of the Sewol ferry tragedy initially had a high level of trust in the President, at least compared to their trust in other political leaders, such as government officials and members of the National Assembly. They believed that the President would at least listen to their grievances; once properly informed, the families believed that she would resolve the issues. In a way, the people's belief in the President reflected the unique characteristics of the Korean political system that has invested the President with enormous powers. Unfortunately, however, President Park Geun-hye's reactions to the bereaved families were so dismissive and even unjust that the families plunged into an even deeper despair. More importantly, their disappointment with the President drove their political struggles into a new phase. Now the focus of their critical consciousness moved beyond corrupt officials, politicians, and the powerful. Rather than focusing on individuals or groups, the families started to think critically about the nation's political system itself.

On April 16, the President did not respond to the accident until 5 pm, though it was revealed that she had received a written report that morning. Various rumors about her absence during the critical time circulated, but little was confirmed. On April 17, President Park visited the indoor gymnasium on Jindo Island to face the wailing and shouting of the families. Badly rattled, President Park promised, "I will rescue them to the last one" and "I will reveal and punish those in charge [of this accident]."[7] On the third day, the commissioner of the Korea Coast Guard briefed the President at the gymnasium on the progress of rescue operations, including information about the involvement of hundreds of divers. The crowd started to jeer even before the commissioner could finish his briefing; they knew what he said was far from what they saw. Just then, a few parents yelled, "That's a lie!" and people started to stir.

[7] *Yonhapnews*, April 17, 2014, "President Park visited Jindo Gymnasium… The families of missing people protest" ("박대통령 진도체육관 찾아 … 실종자 가족 항의"). http://www.yonhapnews.co.kr/politics/2014/04/17/0501000000AKR20140417094554001.HTML.

"Whoever tells a lie must be prepared to resign from office," President Park responded. The statement helped to calm the crowd.[8]

Despite apparent decisiveness, the remarks actually revealed how incompetent and misleading President Park was in responding to the emergency situation at the critical initial stage. Instead of spearheading rescue operations as both the head of administration and the commander-in-chief of the nation's armed forces, she did little to command lower-level officials. In fact, the rescue failures owed much to Coast Guard officials refusing to allow other forces, like the Korean Navy and private marine experts, to join the rescue operation. Instead, the Coast Guard insisted on giving priority to workers of Undine Marine Industries, a private salvage company already under government contract. That process resulted in delays and the ultimate failure to rescue more passengers trapped inside the sunken ferry. This absurd situation occurred because President Park did not grasp the nature of the disaster accurately and failed to take the necessary measures as the leader of the nation. Without the President's authorization, the Korea Coast Guard could not use available manpower and resources beyond office regulations and budget limits.

On the morning of April 20, frustrated by the irresponsibility and incompetence of the officials in charge of the rescue operation, the bereaved families began to march toward President Park's residence, known as the Blue House. Although only a few rubber boats were floating on the sea, the mass media was spreading the false information that the government was carrying out a massive rescue operation. In a blaze of anger, the parents stormed out of the Jindo gymnasium and marched north. "The government is a murderer!" they shouted. "Revive our children!" It was the first collective action by the families. Yet, their march was soon stymied by government forces. The quick reaction of the police was in stark contrast to the government's slow response to the disaster; that contrast hurt the families deeply. For many of the protestors, the experience of standing against the state's armed forces at the Great Jindo Bridge revealed the fundamental character of the state as a repressive apparatus.

---

[8] *Newsis*, 17 April 17, 2014 "President Park, 'Give all of one's strength, otherwise at the risk of one's position'" ("박대통령 '옷 벗을 각오로... 구조전력을'"). http://www.newsis.com/ar_detail/view.html?ar_id=NISX20140417_0012862797&cID=10202&pID=10200.

On May 16, 2014, one month after the tragedy, President Park invited 17 representatives of the bereaved families to the Blue House. In this closed meeting, the families had a chance to tell the President about their sorrows, worries, expectations, and plight; while listening to their heartbreaking stories, she also answered their questions one by one. The conversations covered many issues. One statement by the President was remembered and echoed by many participants in the meeting: "I think that the deep scars on your hearts will start to heal when you have no regrets in the process of finding the truth." For many of the families, President Park seemed to sympathize with the depth of their pain.

That belief was soon shattered. On May 19, three days after the meeting with the families and 34 days after the accident, President Park released the first formal statement on the Sewol ferry disaster.

> The ultimate responsibility for not handling this accident properly belongs to me, the President. I will not allow the noble sacrifice [of the victims] to be in vain but to be a momentum to help the Republic of Korea to be born again. … [I]f necessary, I will form a special prosecution team to disclose the truth in detail and punish [the persons in charge] severely. Moreover, I propose making a special law including a fact-finding commission, which the ruling and the opposition parties and the people will participate in.

The President's statement was nationally televised. Many viewers of the speech, who lacked accurate information about the disaster and the government's response, believed President Park was taking adequate measures and assuming appropriate moral responsibility as the national leader. From the families' perspective, however, it was no more than empty words; it seemed only aimed at the upcoming local elections on June 4. As the elections concluded, that picture changed as suspected. The parliamentary investigation on the disaster was made perfunctorily, revealing almost nothing new. Negotiations among three parties over the Special Act limped along and minimized its operational force. The formal statement of President Park amounted to little more than an exercise in the "politics of language" as well as the language of politics.

As the months passed, the victim families' grief and frustration only grew. From their perspective, the Sewol ferry disaster was still unfolding, and yet Korean society was not changing. On August 22, 2014, the families staged sit-in protests at the Chung-un community service center near the Blue House. Demanding a meeting with the President, Mr Kim Young-Oh, often called "Yoomin's dad" (유민아빠), initiated a hunger strike and visited the Presidential place every day for over six weeks. Although President Park had told the families at the aforementioned May 16 meeting, "I will make every wish of yours fulfilled. Come to see me anytime you want," she now turned deaf to their concerns. President Park would not budge an inch, even when the families made "three steps and one bow (삼보일배)" to deliver 4,850,000 signatures to the Blue House on September 2. A mother who had attended the May meeting sobbed and said:

> When I told [the President], "Please remember our children; please don't make their death in vain," she said, "Sure, I will remember [the children]. Don't worry." This one sentence, "Don't worry," still lingers in my ears. The very sentence she spoke while wiping her tears away with her hand is still lingering in my ears. Yet now she has made a 180-degree turn on her attitude and that makes no sense to me at all. I wonder how the President can cheat on the people like this. We are not asking her to do something for us, but what we want is merely to talk to her. Just one phrase [of consolation], "You have gotten through really tough times" (고생하셨어요) or just one attention would make us feel softly warmed and comforted a bit. Why can't she do such a thing? I am so bitter and resentful. In the past we desired to receive things such as the President's word or consolation, but now we don't. Why? [It is] because at the National Assembly Building she pretended as if she had never met us. Now we will do the same [and ignore her] even if the President is willing to see us.

Through this experience at the Chung-un community service center, the bereaved families were no longer concerned with the President. The attitude of President Park did not change in the winter; nor did it change into the new year. Despite the government's pledge to salvage the Sewol ferry in November 2014, no action was taken for six months. On April 22, 2015, a week prior to the Assembly by-elections (보궐선거), the government announced that the ship would finally be salvaged. To the families, the announcement was one more political manipulation of the disaster. It led to one more stage in a long struggle against the state and the politicians. As that painful journey spanned a year, the families of the Sewol ferry tragedy gradually grew in their political consciousness; their political subjectivities were transformed. During that tragic year, they came to understand that government officials, politicians, nor the

President would help them; the families would have to stand on their own to achieve their goals.

Feelings of Guilt and Life as Devotion to Social Reform

What was the bereaved families' everyday life like prior to the disaster? Were there any characteristics of their lives, especially as workers and parents in Ansan city, that deserve our consideration to better understand their persistent struggles for two years? These are the key questions of the second part of this chapter. In order to answer them, this section focuses on life in Ansan city as part of the broader context of Korean society, the specific characteristics of the families' everyday lives before the accident, and how such characteristics have influenced the families' political awakening and activism.

"People say, they won't live here, they won't live here, and yet they continue to live here—that here is Ansan" (안 산다, 안 산다 하면서 사는 곳이 바로 안산이에요). A pun using homonyms, this is what Ansan residents customarily say when they are asked about their dwelling place. The expression raises many interesting questions: Why do residents dislike living in Ansan? Why do they continue to live in the city? At the very least, the residents of Ansan have contradictory feelings about their community. The expression reveals a bitter self-awareness of living an undesired life but having no other choice.

The conflicted attitude of Ansan citizens is not a simple paradox; it is a historical and political construct. Many people in South Korea are less familiar with the name "Ansan" than they are with "Banwol," the name given by Japanese colonizers. That name remains connected with the Banwol industrial complex of the late 1970s. Banwol was the first planned city in South Korea and a symbol of "developmental dictatorship" (개발독재) under President Park Chung-hee, the father of President Park Geun-hye. As part of economic growth and population distribution projects in the late 1970s, the government planned the city, adopting a method of urban design that considered housing, education, and living environment altogether. However, the construction of the city was, from the beginning, aimed at making the life of residents of Seoul and other metropolitan areas better, rather than to serve Banwol's own citizens. It was built as a satellite city of the Seoul metropolitan area. The city was designed to solve the growing problems of housing shortages and population growth in the capitol of Seoul. Banwol was also used to accommodate all small- and medium-sized businesses and pollution-causing factories that had been located in Seoul and the surrounding metropolitan areas. When the region grew big enough for the status of a municipality in 1986, it discarded the colonial name Banwol for the more historic name of Ansan.

As a planned, state-led industrial complex in which hundreds of small- and medium-sized enterprises and factories are concentrated, Ansan has long served as a settlement for those who have no other jobs or places to go. "No matter how poor the person is, he or she can survive here," Ansan citizens say. The city has provided a refuge for those who were, regardless of their origins, left out during Korea's rapid economic development. Residents included a great number of peasants who had no choice but to leave their lands in Jeolla Province in the 1970s; former coal miners who had no place to go after the mine closures in Gangwon Province in the late 1980s; foreign migrant workers who followed the so-called 3D (dirty, dangerous, and difficult) jobs in the 1990s and suffered from low income and racial discrimination; and numerous heads of households, mostly fathers, who lost their jobs and fell into debt in the aftermath of the IMF financial crisis of the late 1990s (Oh et al. 2008).

While Ansan is also a residence for those who have endured far less painful lives and those who accumulated wealth in the process of industrialization of this area, the social stigma of Ansanians as fugitives, wanderers, criminals, foreigners, or the poor has long defined the city's ethos, even as the negative image has gradually faded over time. In this context, Ansanians have developed strong psychological bonds, a culture of mutual assistance, and an understanding of labor and political rights. At the same time, they have embodied a sense of dissatisfaction and alienation related to their way of life. Ansanians' contradictory feelings about their community are an outcome of Korean modernization. Despite strong pride about many successful stories about civil movements and social reforms in this area, Ansanians have harbored an ambition to escape from their current situations and find a better future.[9]

Indeed, the bereaved parents' self-reflection and repentance for their past lives provide strong motivations for their tireless struggles after the Sewol ferry tragedy. The key content of this self-reflection can be

[9]The history and characteristics of Ansan city are based on interviews with local activists.

summarized as follows: "People are more important than money (사람이 돈보다 *mal*)." The expression sounds like a cliché, but through their painful experiences, the families of the Sewol ferry disaster have realized many people, especially the powerful, are more protective of their money than of other people's lives. They have also witnessed first-hand how materialistic values can lead to the deaths of innocent children. Furthermore, going beyond social critique, the families have regretfully admitted that they had actually lived similar lives, prioritizing money over taking care of those whom they love. They wish that they could have reached this realization before they lost their children. Their regrets were particularly related to three fields: labor, consumerism, and child education.

It is noteworthy that the student victims of the Sewol ferry tragedy were mostly born in 1997 when the IMF cold wave struck South Korea. Like many other heads of households in the country at that time, most fathers of the Sewol ferry disaster struggled financially when their children were born and growing up. Some faced unemployment, some took on extra work during night and weekends, and others confronted bankruptcy and piles of debt. Even if the bereaved fathers are different in terms of age, occupation, and the time of settlement in Ansan city, they all share the sense of "I am sorry" (미안함)[10] to their lost children. They regret not having spent more time with them. One father's confession is especially poignant:

> Before this accident, I lived in order to solve the problem of making a living (먹고 사는 문제). Since solving this problem was far beyond my ability, I rarely went on trips with my daughter. This is why she had so much expectation for this school trip, but then we had this accident. After the accident, I began to think about the problem of living and dying (살고 죽는 문제) instead of the problem of making a living. No matter how

[10] The sense of "I am sorry" is distinguished from the feeling of guilt, though the bereaved parents have experienced both feelings. The feeling of guilt is more directly related to the self-realization th^{a}t they should have acted differently. For example, many parents regret and feel guilty about that they had naively trusted the government and had not told their children to get out of the ship when the accident had occurred. Comparatively, the feeling of "I am sorry" is more related to the family's life condition that is essentially out of control. Many parents, for example, feel sorry about the fact that they were not able to spend more time with their children due to busy daily lives as a wealthless working class.

> economically difficult your life may be, as long as you have people (사람) around, you can still live and talk about your dreams and hopes, even if you have only cheap instant noodles to eat. However, the problem of living and dying is not like that. It is of a totally different kind because without people, everything is meaningless.

In a sense, these fathers were also victims. Their everyday lives were subject to the micro power of the state, which ultimately serves to maintain or reinforce the patriarchal and capitalist order. Gender ideology demands a man takes primary responsibility in bringing food to the table. And if he fails in that responsibility, his ontological value as the patriarch dissipates. Their everyday lives thus logically focused on bringing money to the family. As for people with underprivileged backgrounds, money was also believed to be the only means with which to improve their children's prospects for happiness and success. These fathers, most of whom only had high-school degrees and endured frequent belittlement and discrimination, knew from their own experiences that only those who have an elite university diploma are allowed to enjoy life with comfort and dignity. It was therefore natural for them to yearn for more money and a better chance to send their children to a good college as the primary goal in their lives. However, in so doing, these fathers were complicit with state power even if they did not intend to be. One bereaved father summed up the dilemma: "When I talked to my son, I mostly talked about money. I mean, in order to live on, in order to live happily, money is not everything in a way, but only with money can you go on a trip you want to take and do whatever you want to do."

Indeed, the children's deaths changed everything. Now money acquired an opposite meaning: far from providing the children with a better future, money caused their deaths. If there was one common concern shared by all those who were directly or indirectly responsible for the children's deaths—whether the sailors, the owners and inspectors of the ship, the Coast Guard officers, the government, the media, the politicians, and the President—they prioritized money over the lives of the children. Not only did the families reach critical reflection on those responsible for the children's deaths, but also they came to a self-reflexive conclusion that they too had prioritized money over everything else and been complicit with the micro power of the money-centered society.

Meanwhile, the biggest regret of most mothers of the student victims is that they were too busy to understand the innermost thoughts and feelings of their children. One mother's story epitomizes such feelings of guilt. One day, her family was having breakfast over a little table as usual. Her son abruptly asked, "Can I have a pair of Nike shoes, please? I really want to have them." Due to its famous brand name, Nike shoes are too expensive for a working class family to purchase. That morning, her son obstinately demanded a pair and a peaceful conversation over breakfast devolved into a heated argument. The parents did not understand their son's persistent demand and inconsideration for the family circumstances. The mother scolded, "Do you know how much they cost? Do you know how much parents need to work in order to buy them?" His father finally left his seat spitting out, "Okay, if you really want to have the shoes, I will work overnight again and buy them when I receive overtime pay."

The events of that day have never left her mind since the drowning of her son. At first, she regretted not having bought the shoes sooner when her son was alive. Although the shoes were expensive by their standard of living, they were of no comparison to the value of her son's life. She could have purchased them despite her concerns about the family finances. After her son was found dead, she reflected on the reason why he wanted the shoes so much. Putting herself in her son's place, she realized that the parents cared about the price of the shoes, though that price did not matter much to her son. It was not the price but the shoes that mattered to him. He just wanted to wear the shoes.

After the accident, a number of the families of the Sewol ferry disaster left their jobs. Some could not bear their colleagues' strange looks; some could not follow the work routine due to their activities; and some simply could not find meaning in their work. Unemployment has not merely made their lives economically more difficult; it has also taught them another important lesson:

> It is very strange. When my husband and I were working and earning money before this accident, we always worried about starving to death in the future. Not a single day we did not worry about it. However, as you see, my husband and I, we both have not earned one penny for more than a year, but we have not starved to death yet. People share food with you, and friends bring you to a restaurant. Thanks to the warm-hearted people, we can live on anyhow. Of course, it is shameful that we are still alive though our child is dead.

As shown in these stories, the accident has led the families to change their worldviews, particularly their attitude toward money. They have realized that consumer society compels people to think of everything in terms of monetary value, even though the most important things in life have nothing to do with price tags. They have also recognized that consumer society makes people feel unsatisfied with their present lives and leads them to pursue more materialistic desires. Indeed, the deaths of their children and their struggle against the state have enabled them to see how vulgar capitalist or materialistic culture misleads people and even destroys lives. In the end, the poignant regrets about their past lives formed the bedrock of the families' persistent struggles. Those regrets also saved them from falling for the state's efforts to reconcile with monetary compensation (416 Family 2015).

Like an awareness of labor control and consumer culture, the families' newfound views of child education influenced their political awakening. "Stay still": that was the repeated and haunting order broadcasted throughout the doomed ferry, a call that prevented the students' escape resulting in hundreds of unjustifiable deaths. Of course, it was not the single cause of the disaster. However, the parents of the student victims often questioned themselves: What if their children had not been so obedient as to listen to the superiors' instruction? The parents believed they had trained their children to be submissive to the hierarchical social system. One father was barely able to restrain his anger and self-hatred when he remembered what had happened on April 16. "Don't panic and stay still. Just follow whatever you are asked to do (당황하지 말고 가만히 있어. 그냥 시키는 대로 해)." This was his reply to his daughter when she texted him about the sinking ship. He deplored his attitude that morning. He regretted that he had not told her, "Don't listen to what they say. Just trust in yourself and at a proper time, get out of the ship."

The Sewol ferry accident and the subsequent political struggles taught the families a powerful lesson: in a corrupt and irresponsible society, obedience does not necessarily make life better; it is just as likely to cause misfortune and even death. The families were bitter over the widely circulated ideology of obedience in child education. Through their own experience, they learned that obedience was no more than an effect of power. After this tragedy, the parents of the victims no longer tell their children to obey authority, whether parents, teachers, or other adults. The families have gradually refused to comply with the demands of the state.

Moreover, the bereaved parents have repented that they pressed their children to study hard for college entrance with no regards for what their children really wanted. While reading children's posthumous memos and diaries, parents were surprised to see that their children had actually clearer ideas about their future than they had ever expected. These children also had mature feelings and thoughtful considerations for family members in their plans. The parents' remorse is well expressed in a mother's confession:

> Now I think if I had known his talents earlier, I could have sent him to a different [specialized] school.[11] I feel sorry that if I had not lived in this small city, if I had not worked outside when he was young, I may have nurtured his talent. If I, his mother, had found his talent earlier and not sent him to this school, he would not have died—this thought really hurts me. This morning, I felt guilty about the fact that I had not amply asked my son about what made him the happiest. I just wanted him to become the top in his class and to fulfill what I could not do in my school years.

After all, the parents' guilty feelings that they were complicit in their children's deaths is accompanied by self-critical reflection on lives lived without social participation; their indifference to social issues made them feel vulnerable toward the operations of state power. This self-reflection eventually led them to devote themselves to changing society.

Conclusion

The parents of the Sewol ferry tragedy had been ordinary citizens until their children were drowned. They were mostly from a working class who believed that working hard and having more money would ultimately bring them happiness and respect. The outrage of April 16, 2014 changed their lives and worldviews completely. Based on participant observation and conversations with the bereaved family members,

[11] Due to the high rate of unemployment among university graduates despite hefty university tuition fees, many parents and students in Ansan have recently preferred vocational high schools to general academic ones. Another reason for such a decision is because Ansan has some fine-quality vocational schools. Now vocational high schools have a new name, "specialized high schools" (특성화 고등학교), each of which emphasizes different specialties. Students who want to go this type of school are expected to identify their talents in advance.

this chapter has discussed two crucial factors that triggered them to stop being "passive citizens" and become "resistant subjects." First, the families could not but feel deep indignation and mistrust of the government and political leaders when they observed the post-accident responses and attitudes. They finally decided to stand up on their own to solve painful problems. Second, profound regret and feelings of guilt for their dead children led them to reflect on their past lives and to devote their present and future lives to social reform.

Although it is true that the bereaved families have grown politically through a series of painful experiences (Noh et al. 2015), it would be premature to anticipate positive outcomes of their activities. In fact, there are at least three obstacles that may inhibit their success; the families recognize and sometimes feel anxious about these obstacles.

The first obstacle is that their opponent—the state—is not a weak foe. For any group in today's society, it is almost impossible to challenge, let alone defeat, the state. Even if the current regime is broken, it will be replaced by another regime with essentially similar characteristics. Whoever becomes the next President is unlikely to make fundamental changes; the core issue is not simply the President, but the whole system which allows the political leaders' and the haves' corruption, greed, and lies.

The second obstacle lies with the dissipation of the families' participation and passion for struggle. As expected, more and more people feel tired of fighting a battle that bears little fruit. Some simply wish to start a new life; others have children to take care of. In the beginning, more than 150 families attended the meeting of the Union of the Bereaved Families (유가족협의회). Now its number has decreased to 30 or 40. The remaining bereaved families have witnessed others leaving the group and feel disempowered.

The third obstacle is more complex. Aware of the first and second obstacles, some active family members have begun to think that their struggle should not simply aim at making the state give in to their demands; their struggle should instead build a new form of community that can be relatively free from state power.

Although the Sewol tragedy has ruined the existing relations and way of life among the families, it has also led to the unintended emergence of a community. Confronted with the state's incomprehensible responses, the bereaved families at first gathered to fight against the state. Over three years of being together, these families have gradually realized that other

Fig. 9.2 Mothers of ferry victims sell flowerpots in a local market they organized

bereaved families are more than comrades who have shared the feeling of indignation and guilt; they are friends and even healers who have best comforted their wounded hearts.

Based on the realizations they have obtained in the process of struggle, the families now envisage such a community as a new set of relationships, as well as a new space, in which people work as much as they can enjoy and resist becoming slaves of the micro power of the state. In such a community, the people will assist each other instead of competing against each other. Moreover, children in the community will never be forced to comply with the desires of adults; they will be allowed to choose the happiest life of their own preference. These ideas might sound idealistic, but such ideas come from their lived experiences and painful realizations. They are the outcome of serious soul searching (Fig. 9.2).

Unfortunately, the families have realized that they have few resources to actualize their dreams, however righteous the objective might be. People who may be able to help them—whether through expertise, technology, political power, or financial resources—rarely want to associate with them; just as often, those powerful individuals oppose them. In this social system, the idealism of the families finds little respect or lasting solidarity. On the contrary, the families are in many ways finding themselves cut off from the outside world—just as the Sewol ferry disaster is fading from people's memories.

References

416 Family. (2015). *Will not forget* [잊지않겠습니다]. Seoul: Hanibook.

Noh, M.-W., et al. (2015). *The wind coming from the Paeng-Mok port* [팽목항에서 불어오는 바람]. Seoul: Hyunsilmunwha.

Oh, K. S., et al. (2008). *Ansan in transition: Issue and alternative* [전환기의 안산: 쟁점과 대안]. Paju: Hanul.

Author Biography

Hyeon Jung Lee is Associate Professor in the Department of Anthropology at Seoul National University in Seoul, Korea. Her work as a medical anthropologist focuses on the role of the state and social support in subject formation related to suffering and mental health among peasants in China and the urban poor in Korea. She has authored numerous journal articles and book chapters, including "Fearless Love, Death for Dignity: Female Suicide and Gendered Subjectivity in Rural North China" (2014), "Welfare Programs and the Formation of New Elderly Subjects in South Korea" (2014, in Korean), "Revolution Forgotten: Market Reforms and Left-Behind Women in Rural China" (2014, in Korean), and "Modernization and Women's Fatalistic Suicide in Post-Mao Rural China" (2012).

CHAPTER 10

Epilogue: The Wreck of the Sewol

John Lie

It is a truism of national narratives that certain dates—9/11 in the United States, for example—mark radical ruptures when everything seems to change, and all at once to boot. On April 16, 2014, the South Korean ferry Sewol sank, killing 304 of the 476 passengers on board.[1] What made the tragedy especially painful was that many of the victims were Danwon

[1] For factual information, I have relied on reports in *Tong'a Ilbo* (www.dongA.com), *Chosŏn Ilbo* (www.chosun.com), *Chungang Ilbo* (joongang.joins.com), *Hangyŏre* (www.hani.co.kr), OhmyNews (www.ohmynews.com), and Asahi Shimbun (www.asahi.com). I have also consulted other foreign media coverage. See Seungsook Moon's essay in this volume on the domestic and global reporting on the Sewol ferry sinking.

This is a revised version of the article originally published in *Georgetown Journal of International Affairs*, Vol. 16, No. 2, Summer/Fall 2015. Since most of the names and terms in this essay are widely disseminated in the Anglophone world, I have adopted the most commonly used orthography rather than relying on either the McCune-Reischauer system favored by most non-South Korean Koreanists or the South Korean Ministry of Education system for the body of the text. East Asian names are presented surnames first. I wish to thank Mikyoung Kim and J.J. Suh for inviting me to participate in this volume. This essay is dedicated to the memory of Nancy Abelmann.

J. Lie (✉)
University of California, Berkeley, USA

J.-J. Suh and M. Kim (eds.), *Challenges of Modernization and Governance in South Korea*, DOI 10.1007/978-981-10-4023-8_10

High School students on their school trip. Not only were their youthful lives cut painfully short, but their horrible experiences were communicated more or less instantaneously by social media for all to witness. For months in South Korea, the Sewol tragedy dominated newspaper headlines, evening news, and cyberspace chatter. Reports and revelations, rumors and speculations, alleged cover-ups, and conspiracy theories spiraled into a sea of accusations and counter accusations. Parliamentarians bickered for weeks to agree on committee members to investigate the accident. Meanwhile, South Korean streets became quieter; public displays of merriment were proscribed. School trips—indeed, many trips—were cancelled as South Korean consumption plummeted. As the South Korean writer Kim Young-ha put it: "To outsiders, the Sewol disaster may seem like another tragedy that we will inevitably overcome. But here in South Korea, it feels like the country may never be the same again. It has traumatized our national psyche and undercut our self-image."[2] As tragic as the accident was, however, it remains puzzling why it should be a transformative cultural phenomenon, a moment of national reckoning. My answer is that the wreck of the Sewol suggested the wreck of the South Korean body politic.

The Sinking of the Sewol

Accidents happen; ships sink. Yet what made the capsizing of the South Korean ferry newsworthy was not only the sheer scale of the tragedy (it was widely reported as the worst maritime disaster in South Korean history) but that, as information disseminated, it seemed in retrospect to have been so patently preventable, or at least that the damage could have been minimized. Several of the authors cover the same ground, but let me summarize the basics of the Sewol disaster.

Exact details of the accident remain contested, especially the contingent and conjectural factors immediately prior to the capsizing, but no one disputes many of the conditions that contributed to and exacerbated the death count.[3] First, the Sewol was an antiquated ship with

[2] *New York Times*, May 6, 2014.

[3] The bibliography of the Sewol disaster is already massive. Already in the same year, there had been a barrage of publications; see e.g. Kwak Tong-gi, *Sewŏlho ŭi chinsil*, Seoul: 615, 2014, and Wu Sŏk-hun, *Naeril su ŏnnun pae*, Seoul: Wungjin, 2014. In English, the best summary can be found in Jae-Jung Suh, "The Failure of the South Korean National Security State: The Sewol Tragedy in the Age of Neoliberalism," *The Asia-Pacific Journal*, Vol. 12, Issue 40, No. 1, October 6, 2014. See also the introduction to this volume.

compromised safety features. The shipping company Chonghaejin had bought a nearly 20-year old Japanese ship, which was retrofitted to carry more passengers and cargoes. At the same time, Chonghaejin vitiated her safety features, such as by lowering the amount of ballast water used to stabilize the ship or by failing to anchor adequately the cargoes on board. Old and overloaded, the ship seems in retrospect to have been headed for disaster.

The problem was not merely the hardware. A temporary and inadequately trained crew guided the old, overloaded ferry (19 of 33 crew members were temporary, untrained workers). The criminal disregard for safety training can be gleaned from Chonghaejin's accounting records, which revealed that the company had spent just over $500 in safety training in 2013 when it had spent $230,000 in publicity. After the ship began to turn over, the captain and the crew scurried to rescue themselves and abandoned the passengers. Most catastrophically, many passengers, mostly high-school students, were told to stay put in the cabin rather than evacuating. The reckless disregard of the fundamental ethos of the sea—the responsibility of the captain and crew to ensure passenger safety—rightly attracted a great deal of opprobrium. As reprehensible as the captain and the crew's dastardly behavior was, they have become the scapegoat, as no one else has been tried for the accident. Be that as it may, the software of the Sewol was clearly awry, exacerbating the defective hardware.[4]

Finally, there was the puzzling failure of the rescue operation. The South Korean Coast Guard was not only ineffective in rescuing the passengers but it also proved lax in soliciting, and possibly blocking, help from other agencies, including the South Korean Navy and Coast Guard. Both the US and Japanese offers of assistance were rebuffed. In particular, the National Intelligence Service, which should have spearheaded

[4]There is a temptation to ascribe the accident to South Korean cultural characteristics, such as the pervasive invocation to "hurry" or the common trope to "not to worry" (see e.g. Muroya Katsumi, *Disu izu Koria*, Tokyo: Sankei Shimbun Shuppan, 2014). The 2013 Asiana Airline crash also elicited speculations about the role of culture (as influentially articulated by Malcolm Gladwell about an earlier South Korean accident). See Bob King and Adam Snider, "The Fallacy of Blaming Korean 'Culture' For Asiana Crash," available at http://www.politico.com/story/2013/07/the-fallacy-of-blaming-korean-culture-for-asiana-crash-94053.html. In this regard, the essays by Su-dol Kang and Kyung-Sin Park make distinctive cases for the importance of culture in social analysis. It would be pointless to deny the salience of culture, but it seems equally problematic to overstress it, however.

the rescue effort in a major accident, remained immobile and impotent. President Park Geun-hye was either absent or uninformed. Clearly, the state and its apparatuses performed poorly in the face of a major national disaster.

As outsiders are wont to observe, ships sink with some frequency. Furthermore, the 2012 capsizing of the Italian luxury vessel Costa Concordia also featured a captain who abandoned the ship prematurely before every passenger had been evacuated.[5] Yet even a cursory comparison suggests a world of differences. The Costa Concordia was relatively new and reasonably well maintained. In spite of the captain's premature disembarkation, many of the passengers had been safely evacuated. Finally, the rescue effort was relatively successful, ultimately resulting in only 32 deaths of the 4229 passengers on board (there was also a comparable accident of a Japanese ship five years before the Sewol tragedy that resulted in no fatalities).

The more the public discovered about the particulars of the Sewol disaster, the more the public could not shake off the sense that rather than being a mere, unfortunate accident, the sinking of the ship was a preventable disaster: a concatenation of human follies and a matter of grave social injustice. Bureaucratic and political cover-ups exacerbated societal anger and acrimonious debate. These include inconclusive explanations of the whereabouts of President Park on that fateful day to the possible malfeasance on the part of the National Intelligence Agency. Beyond accusations and counter-accusations lurked the larger meaning of the preventable disaster. The Sewol tragedy struck many South Koreans as a terrible symptom of an ailing and aimless body politic. The wreck of the Sewol seemed to prognosticate the impending wreck of South Korea.

The State of the Wreck, the Wreck of the State

The sinking of the Sewol suggested the exhaustion of the South Korean model of political economy. Non–South Korean scholars may dispute the primary institutions and forces of the contemporary South Korean economy but few South Koreans would disagree on the primacy of the state and the chaebol (big business). The dominant, received narrative of

[5] See the series of articles in *The Guardian*, available at http://www.theguardian.com/world/costa-concordia.

South Korean economic dynamism inevitably highlights the role of the powerful, developmental state and large business conglomerates, such as Samsung and Hyundai.[6] The sheer rapidity of industrialization and associated social transformations—often summarized as "compressed," and generating manifold contradictions—is elaborated in the editors' introductory essay, as well as in chapters by Yoonkyung Lee, Kyung-Sin Park, and J.J. Suh. There is a great deal of disagreement on the exact nature of South Korean state power and its relationship to big business: is it a neoliberal state (Lee), a hierarchical and paternalistic state (Park), or a state with an uneven exercise of state power (Suh)? How much has it changed since democratization? Not so much, as Su-dol Kang seems to suggest? Or is it in potential danger of backsliding to the authoritarian past, as Taehyun Nam argues?

Contemporary South Koreans, not just scholars within and without South Korea, disagree on the desirability of the state- and the chaebol-led economic model. One of the fundamental cleavages of South Korean politics aligns those who argue that the model is basically sound, believing it to be based on meritocracy and expertise and an engine of economic growth, and those who find it problematic, observing that it is entwined in corruption and collusion and generates inequality and poverty. Whereas Kim Dae-jung's and Roh Moo-hyun's presidencies sought progressive policies based on the latter belief, their successors Lee Myung-bak and Park Geun-hye have largely espoused the former.[7]

The black-and-white portrait of South Korean political divergence occludes the growing and gnawing consensus on the exhaustion of the South Korean political-economic model. There had been a period of serious reconsideration almost two decades ago in the aftermath of what in South Korea is remembered as the 1997 IMF crisis (the aftermath of the 1996 Asian financial crisis). Three decades of uninterrupted and unprecedented economic growth came to a halt, spawning massive unemployment and shattering the myth of an ever-improving economy. It is precisely in this context that the progressive Kim Dae-jung became the president, seeking to temper the developmental state (by enhancing

[6] See e.g. John Lie, *Han Unbound: The Political Economy of South Korea*, Stanford: Stanford University Press, 1998.

[7] For a nuanced empirical examination, see Ōnishi Yutaka, *Senshinkoku-Kankoku no yūutsu*, Tokyo: Chūō Kōronshinsha, 2014.

welfare functions) and the chaebol (by instituting anti-oligopolistic measures). By then the criticism of government–business collusion (elucidated in J.J. Suh's chapter) and endemic corruption (illuminated in Jong-sung You and Younmin Park's chapter) was widespread, becoming something of societal commonsense. Lee Myung-bak, who became president in 2008, had been the president of Hyundai Engineering and was widely perceived as restoring the tattered relationship between the state and the chaebol, which from a different perspective was a restoration of the cozy, collusive, and corrupt state–business nexus. In this regard, Taehyun Nam's chapter questions the extent of democratic consolidation in South Korea. Although President Park broadly shares Lee's politics and in particular the faith in the political-economic model that is usually attributed to her father, it would be misleading to neglect her concessions to the opposition. That is, she has repeatedly acknowledged not only the severity of economic inequality and the inadequacy of welfare provisions but also the widespread concern over chaebol power and the charges of crony capitalism: patterns of collusion and corruption that sustain state–business relations. Yet Nam is right to worry about the "monarchal" presidency, with its excessive authority, and the slide back into authoritarianism. Furthermore, Su-dol Kang's "addictive organization theory" points to the systematic and persistence effect of compressed development on the organizational culture of South Korean government and corporations, which have remained remarkable robust in their dysfunctions, including corruption.

Investigations of the Sewol tragedy brought many of the concerns with the state of the South Korean political economy to the surface. The broadly neoliberal orientation of recent South Korean regimes, as highlighted in Yoonkyung Lee's chapter, loosened regulatory measures. Not only had the Lee regime in its effort to promote business-friendly policy extended the permissible life span of a ship from 20 years to 30 (thereby allowing the Sewol to operate and ultimate to capsize), it had also vitiated many of the safety measures, thereby promoting profit motive over consumer protection. One telling indicator of lax government inspections that emerged was the systematic underreporting of cargo load: at the time of the accident, the Sewol underreported its load by 500 tons. Just before the fateful voyage, moreover, the inspection was completed in under an hour. Beyond generic pro-business policies—lax regulations and inspections, for example—the Sewol's owner Chonghaejin turned out to be the beneficiary of government largesse, benefitting from a combination of debt forgiveness and new loans. Furthermore, speculations have mounted on the collusion—and corrupt dealings—between government agencies and Chonghaejin. Jong-sung You and Younmin Park's chapter argues that, in spite of the secular decline in overall corruption, "regulatory capture" is rife: incentives and incidents of corruption remain high in regulatory agency and targeted industry. Indeed, allegations of kickback extended to school officials. As J.J. Suh argues, the ineffective rescue operation cannot be told apart from the collusion of market interests with government agencies. All the unsavory facts—and unproven accusations—recall the bad old days of South Korean capitalism that was cemented by bribes in the name of personal relationships.[8]

There is an ironic overlay to the Sewol disaster. One of the early triumphs of South Korean capitalism was the growth of Hyundai Shipbuilding. Harking back to the late sixteenth-century naval commander Yi Sun-sin's "turtle boats" that turned back Japanese invasion—the 2014 movie about Yi, The Admiral [Myeongnang in Korean], remains the most popular South Korean film of all time—many South Koreans have taken great pride in technological prowess in general and in shipbuilding in particular.[9] In spite of the widely trumpeted claim to be the leading shipbuilding nation, however, the Sewol demonstrated the disjuncture between export success and domestic failure. Many South Koreans are keenly aware that the government's longstanding export-promotion strategy has helped big business but has hurt South Korean consumers. For example, it is often cheaper to buy a South Korean product abroad than at home. The same logic of export and profit first, South Koreans' welfare second, seemed to operate in the case of the Sewol disaster. Rather than boarding a state-of-the-art luxury liner, South Koreans were on an antiquated, non-South Korean ship sailing to their death.

Beyond a political-economic system that favors big business over people lies a societal suspicion of inequality and injustice. The "nut rage incident"—the 2015 event in which a chaebol daughter-executive forced flight attendants to grovel in front of her and pilots to divert the plane

---

[8] See especially Kwak, Chap. 6. See in general Jong-Sung You, *Democracy, Inequality and Corruption*, Cambridge: Cambridge University Press, 2015, especially Chap. 3, in addition to the chapter by You and Park in this volume.

[9] A more disturbing interpretation was to underscore the shortcomings of the South Korean shipping industry, particularly its inability to manufacture high-end vessels.

and its plan at her whim—emphatically showed that the gap between the economic elite and the masses appears unfathomable. Yet extreme economic concentration is increasingly seen not so much to generate economic dynamism as to cause poverty and inequality. In the early 2010s, the total sale of the top 10 chaebols accounted for nearly 80% of South Korean GDP. Rather than celebrating them as the engine of economic growth, chaebol dominance has elicited negative sentiments; well over two-thirds routinely express anti-chaebol opinions. The developmental state in turn has proven to be a laggard in terms of providing welfare benefits or old-age pension in a society that is rapidly aging. Symptomatically, Danwon High School—the school that most Sewol victims attended—is located in the predominantly working-class city of Ansan. Here Hyun-ok Lee's chapter casts light on a dimly recognized reality of South Korean life: the influx of immigrants and the making of a multiethnic society.[10] Focusing on Vietnamese-Korean victims of the ferry disaster, she points to the "double exclusion" of ethnicity and class in South Korea. The Sewol was yet another stark reminder that the South Korean political economy appears to be exacerbating inequality and poverty, empowering the chaebol while disempowering the people.

The Sewol disaster suggested that South Korean state–chaebol capitalism may not be serving South Korean people. The sinking of the Sewol, then, can be understood as an ominous sign on the fate of the South Korean economy guided by the state and the chaebol.

A Society Adrift

Nevertheless, can the possible exhaustion of the South Korean political-economic model generate so much national anguish? Disagreements on the desirable level of inequality or on the optimal nature of government intervention in the economy are universal. To be sure, when South Koreans talk about the two Koreas, they are just as likely to be thinking of deep divides within South Korea as the split between North and South Korea. The obvious flash point is the assessment of President Park Chung-hee, the dominant figure from the 1960s and 1970s: some view

---

[10]See John Lie, ed., *Multiethnic Korea? Multiculturalis, Migration, and Peoplehood Diversity in Contemporary South Korea*, Berkeley: Institute of East Asian Studies, University of California, Berkeley, 2015. As noted, Hyun-ok Lee's chapter does an important job of recuperating the multiethnic dimension of the Sewol disaster.

him as the architect and even the savior of contemporary South Korea; others remember him as a tyrant. Yet as the ferocity of the Blue States–Red States split in the United States suggests, intense intra-national rivalries are surely common enough in post–Cold War OECD countries. It is not just the exhaustion of the political-economic model that disturbs many South Koreans but perhaps more importantly the very constitution of the nation. The Sewol tragedy struck South Korea precisely when a majority saw South Korean society as being aimless and adrift, and experienced the sinking feeling of being in an ailing, possibly dying, body politic.

The claim of South Korean malaise would strike most casual foreign observers as extremely bizarre. After all, South Korea is a showcase for emerging economies that it is possible to become an advanced industrial powerhouse. As the 2013 global pop music sensation "Gangnam Style"—and in general the Korean Wave, the export-oriented South Korean popular culture industry—exemplifies, South Koreans seem to have become adept at not just producing ships and chips, smart TVs and smartphones, but also cool and creative cultural products.[11] It reveals a great deal about South Korea that its showcase products these days are cosmetic and popular entertainment. Should not South Koreans—at least the comfortable classes of the country—be rejoicing in their successes?

As in many countries, what attracts outsiders may not be what obsesses insiders. For example, most news items about South Korea outside of South Korea involve North Korea. Every North Korean missile launch generates headlines around the world, but nonchalance or insouciance is the modal response among South Koreans, who have heard the proverbial wolf cry too many times. As much as the sitting president may trumpet the triumphs of South Korean economy or culture in the world, what animates domestic debates are economic, social, and cultural problems. Seungsook Moon's chapter shows the divergent representation of the disaster between South Korea and the rest of the world. Indeed, and paradoxically perhaps, the sense of crisis is no less felt among reactionary or conservative supporters of President Park. Societal malaise is pervasive.

---

[11] See John Lie, *K-pop: Popular Music, Cultural Amnesia, and Economic Innovation in South Korea*, Oakland: University of California Press, 2015.

Certainly, South Korea often ranks first among OECD countries in measures that are not particularly impressive. South Korea frequently features the lowest fertility rate and the highest suicide rate: the net effect is to present the country as dying. Indeed, South Korea is rapidly aging while youth unemployment remains extremely high. It is possible to contextualize these dispiriting statistics in the larger canvas of South Korean economic dynamism. Yet it is precisely rapid economic growth that has unshackled the population from its historic hometowns and cultural traditions. While gerontocratic norms prevailed a generation ago, the rapidly aging society faces the recalcitrant reality of underdeveloped welfare and pension systems, precisely when traditional social networks and protections have weakened. The nuclear family as the potential haven in the heartless world also turns out to be deeply fraught, as evinced by the high and accelerating divorce rate and the huge generational gap in experience and expectation (though Hyeong Jung Lee's chapter points to the persistence of the family as a primary institution in South Korea). Hitherto unimaginable social phenomena, such as homelessness, vividly demonstrate the inescapable reality of poverty but even more the fraying of the social fabric. Once a country noted for its insistent homogeneity—the pronouns "we" and "our" precede any discussion about South Korean characteristics—the social divides are manifest and deep: not just the left–right split or income inequality I mentioned earlier but the chasm between the old and the young, the city and the country, the women and the men. Consider only the raw statistics from the 2012 presidential election: while over two-thirds of those under the age of 25 voted for Moon Jae-in, over 80% of those 65 or older voted for Park Geun-hye.

The Sewol tragedy is the proverbial nail in the coffin to the myth of the happy and homogeneous South Korean nation. Whereas a rightwing social critic questioned why the high-school students on the Sewol were traveling to Jeju Island and not to the landlocked Gyeongju (implying an unnecessary luxury), a parent of one of the victims exclaimed that things would have been different had they hailed from the prosperous Gangnam and not the proletarian (and multiethnic) Ansan.[12] Indeed, the bereaved families demonstrate that the relatively powerless in South

[12] Ansan is a modest, undistinguished city where Danwon High School is located. On the exchange, see Wu, Chap. 4.

Korea are far from being passive and consenting to the status quo. Hyeong Jung Lee's chapter illuminates the civil resistance and the making of "active subjects" by the bereaved families seeking restitution and justice.

Serious inequality in the context of undeniable enrichment accentuates the pain of the accident. Whereas few spoke critically about the hazardous working conditions of early South Korean industrialization, nearly everyone is conscious of the differential risks that the rich and the poor face today. The putative unity of the people, endlessly urged by South Korean political leaders—one of the few points of agreement between the right and the left in post-Liberation South Korea—has shattered, perhaps beyond repair. The rightwing Korean Parents' Federation went so far as to throw food at the victims' parents who were on hunger strike to seek truth and justice about the Sewol tragedy. The loss of national unity is plain and searing. Public debates and political disagreements are not just pointed but barbed and even barbaric.[13]

More disturbingly, the very future of South Korea seems to be in question. It is a trite truth that the children are the future of any society but the depressing spectacle of the dying students on the Sewol symbolized more than unnecessary and untimely deaths. At the risk of seeming to blame the victims, it is incontrovertible that the students obediently followed an insane order to stay in the cabin while the ship was clearly capsizing. Kyung-Sin Park's essay stresses the legal culture of hierarchy and authority underscored in turn by state paternalism, which contributes to unreflexive obedience, and Seungsook Moon's chapter highlights the neoliberal, consumer capitalist culture that pacifies independent and critical thinking among the youth. The passivity of the children—the South Korean discourse inevitably describe high-school students as "children"—is also a product of the educational system almost universally believed to be dysfunctional. The ostensibly meritocratic system of examinations determines the life chances of almost every South Korean. An extremely common adaption is to enter examination-preparation centers (cram schools or *hagwon*), which begins as early as elementary school for children who may attend a series of cram schools until midnight in some cases. The *hagwon* heaven has made some teachers superstars earning millions of dollars but the systematic consequence

[13] See e.g. Yu Sun-ha, *Tangsindŭl ŭi Ilbon*, Seoul: Munidang, 2014, Chap. 1.

is widely feared to be conformity (looking for the right answers) and the suppression of creativity and even humanness. The irrepressible suspicion is that the educational system—the heart and soul of the South Korean economic miracle—may have spawned a new generation of passive and obedient youths who will quietly and obediently sink to their death. It is not a stretch to see the dysfunctional educational system as potentially destroying the future of South Korea, as much as the educational system was often celebrated as a major source of South Korean economic dynamism. Indeed, it is almost impossible to find anyone in South Korea who will defend the educational system outright; yet, in the same breath, almost everyone argues that nothing can be done to reform or overthrow the current, dysfunctional system.

The Sewol tragedy laid bare the impoverished spirit of contemporary South Korea. There are older South Koreans who rejoice in the nostalgia of the golden years of economic growth but even they do not assert that all is well or that the future is secure. In any case, few take solace in traditional culture when it is under constant assault. It is not only the political-economic model or the educational system that has been questioned. Christianity, perhaps the leading religion in contemporary South Korea, has gathered a great deal of negative publicity recently, and the Sewol tragedy merely added another scandalous note. Yoo Byung-eun, the billionaire owner of the Sewol, was also a leader of a Christian sect. Reports of the tragedy cast unfortunate light on Yoo's unsavory religious and business practices, including his opulent lifestyle. Cynical manipulation of believers and the hypocritical behavior of manipulators constitute a depressing spectacle. What can and what should South Koreans believe in? Deep social divides in any case make impossible a sense of common purpose, despite decades of hyper-nationalist propaganda.

The sinking of the Sewol provided a convenient receptacle to collect and concentrate all the disparate discourses of South Korean problems and dysfunctions. And it becomes difficult to resist the conclusion that contemporary South Korea is ailing, anomic, and adrift. Who will rescue the sinking South Korean society?

Conclusion

The sinking of the Sewol serves as a turning point for South Korea. President Park said so herself. It is widely recognized as a crisis: the exhaustion of not only the political-economic model that fueled South

Korean industrialization for the past half-century but also national social cohesion—in spite or even because of conflicts and contradictions—that endowed it with meaning and significance. The specter of the Sewol capsizing presents the disturbing imaginary of the sinking body politic. It should not be surprising that, for many South Koreans, April 16, 2014 proved to be such a resonant date.

It remains unclear what will come of the Sewol tragedy, however. Distressingly, the Park presidency seems intent on curbing dissent and suppressing freedom. If the era of Park *père* were repressive, it was at least replete with meaning and significance and economic growth occurred. The current conjuncture of Park *fille*, in contrast, exists in a clean and well-lit but ultimately meaningless world where the political economy that her father had shaped seems increasingly ineffectual. In seeking rapprochement with China, is it possible that the state–chaebol coalition is departing from the western norm of democratic capitalism to emulate the authoritarian-capitalist model of China?[14] The once-formidable progressive coalition is divided and more disturbingly reactive and revanchist. What is clear is that the status quo is utterly inadequate but a reasonable future—hope—seems unavailable or out of reach. Although it would be wrong and foolish to underestimate the resistance and opposition to the status quo, the present and the future look bleak to many South Koreans. Both the conservatives and the progressives primarily look to the glory days of the past. "Hope had grown grey hairs,/Hope had mourning on ..." The spiritual crisis of the South Korean nation is the moral of the Sewol tragedy.

References

Kwak Tong-gi, Sewŏlho ŭi chinsil. 2014. Seoul: 615.

Lie, J. (1998). *Han unbound The political economy of South Korea*. Stanford: Stanford University Press.

Lie, J. (Ed.). (2015a). *Multiethnic Korea? Multiculturalis, migration, and peoplehood diversity in contemporary South Korea*. Berkeley: Institute of East Asian Studies, University of California, Berkeley.

Lie, J. (2015b). *K-pop: Popular music, cultural amnesia, and economic innovation in South Korea*. Oakland: University of California Press.

Muroya Katsumi, Disuizu Koria. 2014. Tokyo: Sankei Shimbun Shuppan.

[14] Cf. Diana Pinto, *Israel Has Moved*, Cambridge: Harvard University Press, 2013.

Pinto, D. (2013). *Israel has moved*. Cambridge: Harvard University Press.
Suh, J.-J. (2014). The failure of the south korean national security state: The sewol tragedy in the age of neoliberalism. *The Asia-Pacific Journal, 12*(40), No. 1, October 6, 2014.
Sun-ha, Y. (2014). *Tangsindŭl ŭi Ilbon*. Seoul: Munidang.
You, J.-S. (2015). *Democracy, inequality and corruption*. Cambridge: Cambridge University Press.
Yutaka, Ō. (2014). *Senshinkoku-Kankoku no yūutsu*. Tokyo: Chūō Kōronshinsha.

Author Biography

John Lie teaches social theory at the University of California, Berkeley, USA. His most recent book is *K-pop: Popular Music, Cultural Amnesia, and Economic Innovation in South Korea* (University of California Press, 2015).

Index

J.-J. Suh and M. Kim (eds.), *Challenges of Modernization and Governance in South Korea*, DOI 10.1007/978-981-10-4023-8

Kolibri 360 GmbH
Ferdinand-Jühlke-Straße 7
99095 Erfurt, Deutschland
produktsicherheit@kolibri360.de